Chemical Ecology

Series Editor
Françoise Gaill

Chemical Ecology

Edited by
Anne-Geneviève Bagnères
Martine Hossaert-Mckey

iSTE

WILEY

First published 2016 in Great Britain and the United States by ISTE Ltd and John Wiley & Sons, Inc.

ISTE Ltd
27-37 St George's Road
London SW19 4EU
UK

www.iste.co.uk

John Wiley & Sons, Inc.
111 River Street
Hoboken, NJ 07030
USA

www.wiley.com

Library of Congress Control Number: 2016945010

British Library Cataloguing-in-Publication Data
A CIP record for this book is available from the British Library
ISBN 978-1-84821-924-3

Contents

Foreword

The consequences of the global changes affecting our planet are not limited to climate change or to changes in composition of the atmosphere. In perturbing all ecosystems, and thereby the species that constitute them, these changes affect all the dynamics of life. These effects in turn have impacts, notably on interactions between species or between individuals of a single species, interactions that are indispensable for the maintenance of communities and ecosystems, in which chemical mediation plays a predominant role.

Chemical ecology can be defined as the art of decrypting this invisible, impalpable chemical mediation that permits living organisms to communicate among themselves within their environment. A great diversity of molecules, ranging from very simple compounds to highly complex mixtures, is involved in organisms' perception of the environment, in communication between individuals, and in the defense mechanisms that have evolved in those interactions that are antagonistic. This scientific domain has successfully reconciled ecology and chemistry, a *tour de force* that has required a resolutely interdisciplinary approach.

Chemical ecology also provides us with a framework to better interpret, extend and use our knowledge about the diversity of natural substances. Researchers are beginning to understand the roles of these compounds in processes of communication in a diverse range of habitats, both aquatic and terrestrial, and between organisms in all the kingdoms of life: animals, plants, fungi, bacteria and archaea. Beyond such fundamental advances in knowledge, chemical ecology is a source of inspiration for new biosourced applications and helps us conceive of the future ecotechnologies that will be necessary for the resolution of a number of environmental problems.

The research in chemical ecology discussed in this book has been conducted by an internationally recognized, dynamic and original community of French scientists. Chemical ecology has been clearly recognized in recent years as a field that promises to contribute pioneering research, situated at the crossroads of multiple competences shared not only by ecologists and chemists, but also by physiologists, biochemists, ethologists, and ethnologists, among others.

This book will allow the reader to explore the myriad facets of the language of molecules that unites biodiversity and chemodiversity, and thereby discover a new dimension of the living world.

Stéphanie THIÉBAULT
Françoise GAILL
Institute of Ecology and Environment CNRS
June 2016

Introduction

The association of the two terms "ecology" and "chemistry" has recently become evident for researchers, biologists and chemists alike, working at the interface of biology and chemistry. Chemical ecology is now an entire area of research. It is a recent discipline, born during the 1970s/80s, and its development was associated with major progress in analytical chemistry during the same period. This discipline has greatly deepened our understanding of semiochemicals emitted by microorganisms, plants and animals.

To survive and adapt, all living things, from the simplest to the most complex, must intercept the information emitted in their perimeter of perception. The majority of living species communicate among themselves by molecules and chemical signals that we may term "chemical mediators". In effect, any ecosystem is a dynamic assembly promoted by interactions that are, therefore, essentially founded on trophic exchanges such as molecular exchanges, involving complex substances that often transmit simple messages.

The chemical language, using semiochemicals much like words, is, in nature, a kind of universal language and appears to be indispensable for maintaining terrestrial and aquatic ecosystems. Chemical communication is by far the most frequently used mode of communication in the living world.

In an attempt to understand this language of nature, ecologists and chemists are confronted with the complexity and the creativity of organisms. Studying an ecosystem – its structure and functioning, the interactions of

---

Introduction written by Anne-Geneviève BAGNÈRES and Martine HOSSAERT-MCKEY.

organisms within it, among themselves and with their physico-chemical environment – requires a multi- and pluridisciplinary approach, an approach indispensable in chemical ecology, which is a natural interface between the two sciences.

An increasingly large amount of data on organisms and the chemicals that mediate interactions between them, in both terrestrial and aquatic environments, continually reinforce our understanding of the biodiversity and chemobiodiversity of living things. The most innovative aspects are linked to the evolution of the species and the mediation of complex interactions in their multitrophic environment, considering each species as an integral part of a community, not as an entity unto itself (Chapter 1). Characterized semiochemicals, either attractants or repellents, selected over thousands of years of evolution and co-evolution for their efficiency, generally have very specific effects on the target organism. Interactions between organisms involve multiple scales; research on chemical ecology therefore relies on a wide range of experimental approaches. Furthermore, given the current loss of biodiversity and ongoing climate change, it is important to understand the functioning of ecosystems and the interaction between their microbial, plant and animal components before considering the effects of human disturbances on these ecosystems (Chapter 2). Work on sociality has allowed the elucidation of a complex evolutionary history of chemical communication in animal behavior, particularly in social species that we think of as microsmatic (i.e. having limited olfactory sense) such as human and non-human primates. In these species, the chemical composition of body odor can reflect individual characteristics. In addition, the use of natural substances by animals for self-medication, which has been shown in arthropods and vertebrates, including non-human primates as well as humans, emerges as an important evolutionary theme. Semiochemicals can, therefore, be considered as a central element of the organization of most animal societies (Chapter 3). Likewise, recent advances in the chemical ecology of the microscopic living world, a theme that was long largely neglected, have in effect modified an overly simplified image of interactions. Microorganisms – prokaryotes (bacteria, cyanobacteria, archaea) or eukaryotes (fungi, protists) – live in communities where intense competition occurs. In response to particular environmental constraints, these microorganisms produce an entire arsenal of molecules. Understanding the mechanisms by which these molecules are produced, and their effects on other organisms, is indispensable to the understanding of the interactions in microbial communities. The study of how microorganisms adapt in sometimes extremely hostile environments often has applications in the field of biotechnology (Chapter 4). The interactions between components of ecosystems can be disturbed by human activities. The active biological and chemical interactions of components in and with the elements of soil, air or water are called ecogeochemical. Ecogeochemistry thus proposes to analyze, by integrative approaches, the complexity of ecological systems and the mechanisms by which the biotic and abiotic components of the ecosystem interact. It complements the classic "biogeochemical" approach of functional ecology by addressing in a single conceptual context the organisms and components of their abiotic environment, particularly the chemical compounds in interaction with these organisms (Chapter 5). For several years, chemical ecology has benefited from the progress achieved in genomics, transcriptomics, proteomics and metabolomics; chemical ecology has thus entered the era of "omics". "Omics" regularly lead to new tools that are very useful for shedding new light on evolutionary mechanisms. "Omic" approaches in chemical ecology vary greatly and are based on a range of biological models, from the simplest to the most complex (Chapter 6). Metabolomics is the most recent of the "omic" sciences. Metabolomics can be applied without an *a priori* approach, aiming to analyze the largest possible portion of the metabolome. It can also be applied in *a priori* approaches targeting a family of metabolites that belong to a particular path of biosynthesis. Metabolomics provides essential information to clarify the key roles played by semiochemicals in the interactions between organisms and their environment, and the mechanisms regulating these interactions. The increasingly powerful analytical, mathematical and statistical tools made available to biologists and chemists thus enable the consideration of increasingly detailed characterization of metabolomes (Chapter 7). The characterization of mediators by increasingly perfected chemical tools, and the new techniques of genome sequencing, have together allowed all these innovative approaches to contribute to better understanding of the living world and its language. Improvements in instrumentation, with gains in sensitivity and resolution, have made it possible to obtain increasingly precise and detailed analyses of primary or secondary metabolites. These approaches generate masses of data, making indispensable automatic comparison with online databases (Chapter 8). The characterization of a chemical mediator of ecological interactions can lead to multiple applications in the fields of applied research such as medicinal chemistry, pharmacology and phytopharmacy. Characterization of the biological target of a semiochemical can lead to the discovery of new biological receptors. In certain cases, nature can adapt to the presence of high levels of pollutants. In ecosystems seriously affected by pollutants, a combined study of the chemistry and ecology of the plants that are able to develop despite pollution can lead to the development of procedures to decontaminate soils, purify water or air (phytoremediation) and restore ecosystem functioning (green engineering) (Chapter 9). At the end of the book (Conclusion), we address the questions that remain unanswered in this constantly changing discipline.

The scientists, including biologists, ecologists, biochemists, chemists and biostatisticians, who have contributed to this book are interested in both continental and marine environments, both temperate and tropical ecosystems, and in living things ranging from microorganisms to mammals, all of which are covered in their analyses of chemical ecology and of its perspectives. The work presented herein illustrates the most advanced and varied aspects of this rapidly expanding discipline. Compared with other books available on related themes, which for the most part deal with relatively simple systems, covering pairwise or tritrophic interactions and comprising a small number of model organisms and semiochemicals, our book offers a holistic vision of chemical ecology.

As a final remark, we wish to pay homage to Murray S. Blum, who just recently passed away and who was one of the first to demonstrate the importance of chemical mediation in the living world. Among other things, Murray brought the notion of parsimony into chemical ecology, along with his smile, which he distributed without parsimony.

1

Biodiversity and Chemical Mediation

Chemical mediation is a widely used mode of communication that contributes significantly to the organization and functioning of biodiversity. Identifying and classifying species are prerequisites for the study of biodiversity, and as with other morphological or molecular characteristics, the study of traits of organisms related to the production of semiochemicals is regularly used today in integrative taxonomy. One of the important facets of chemical ecology is the study of communication mediated by organic compounds (volatile or not) in the same species (reproduction, meeting sexual partners, etc.) and between species (pollination, predation, parasitism, etc.) with honest or deceptive signals. Identifying the compounds emitted, and understanding their modes of action and their roles in the interactions between individuals and between species, are the objectives of this young pluridisciplinary science, which aims to discover this hidden language of nature [PIC 06, RAG 08]. This interspecies chemical mediation sometimes also allows the creation of complex interactive networks, structuring biodiversity around certain organisms, which often play keystone roles in ecosystems [HOS 10, IVA 11a, IVA 11b]. Ephemeral or stable, from attractive to repulsive, this communication is based on an infinite multitude of combinations of organic compounds, where the game for each species consists of emitting, detecting or even masking a scent.

Chapter written by Bertrand SCHATZ, Doyle MCKEY and Thierry PÉREZ.

1.1. Systematic and integrative taxonomy from chemical ecology

A good understanding of biodiversity is a prerequisite for numerous disciplines, such as biology, ecology and even the study of chemical communication between different species. Systematics is the science devoted to the discovery, the interpretation and the classification of biological diversity. This term designates both the methods implemented and the results of their application, and can, therefore, lead to the "classification of living things" in general. Systematics includes taxonomy, which describes living organisms based on their characteristics, most often morphological and/or molecular (sequencing of DNA or RNA is widely used, notably to characterize the infinitely small), and groups these organisms into taxa. Because it is crucial for analyzing and conserving biodiversity so that each taxon has a name, and a specific name never designates multiple taxa, the taxonomic descriptions must be as precise and detailed as possible, integrating complementary types of characteristics. Today, this integrative taxonomy is considered to be the most rigorous approach in systematics because it integrates all the taxonomic, morphological and ecological traits for which scientific information exists to characterize the taxa considered. This method is also the most adaptable, since a small number of the most important traits in the ecology of species concerned may be selected to determine whether two sets of individuals belong to the same species.

Like the morphological or molecular characteristics of an organism, its chemical composition – i.e. its *metabolome* – can be used as a characteristic (or a set of characteristics) in taxonomy or systematics. Chemotaxonomy (also called chemosystematics) seeks to understand the relationship between the chemical composition of organisms, their taxonomic identity and their systematic classification. The metabolome can be studied as a signature of evolution, and the metabolomic revolution is transforming chemosystematics by making it possible to quickly compare a large number of such chemical signatures (Figure 1.1) [BAG 10a, CAR 12]. The analysis of portions of the metabolome also provides classifications similar to those supplied by molecular systematics based on the analysis of portions of the genome. Thus, it provides support for hypothetical classifications. Chemotaxonomy can also be used to discriminate "sister" species, notably for difficult taxonomic groups in which closely related species are often "cryptic", owing to the small size of organisms, the absence of morphological variability or, on the contrary, excessive variability.

In the case of sponges of the class *Homoscleromorpha*, the absence of variation in morphological characteristics classically used in sponge systematics (skeletal spicules) led researchers to search for other informative characteristics. Discovery of variability in up to six different types of characteristics has dismantled the "myth of the cosmopolitan species *Oscarella lobularis*", and 15 species of this genus have now been described [IVA 11a, IVA 11b]. Metabolomic approaches combined with traditional and molecular systematics have recently allowed the proposal of a new systematic classification of *Homoscleromorpha* sponges, and integrative taxonomy is now used to describe many other species of this class of sponges [BOU 14, RUI 14, CAC 15] (see Chapter 7).

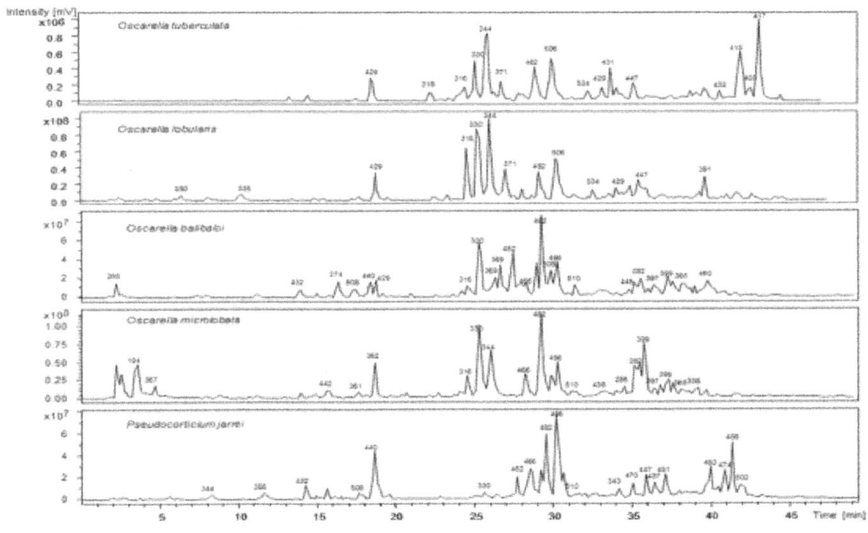

Figure 1.1. *Metabolomic fingerprints showing interspecific variability of the chemical signal emitted by different sponges of the family Oscarellidae, Class Homoscleromorpha: HPLC-ESI (þ) MS (BPC) with indications of m/z values above the peaks of the major compounds (from [CAR 12])*

In the case of Mediterranean orchids, the morphological similarity of the described species sometimes makes their recognition difficult, notably by stakeholders in conservation efforts. Thus, in 2010, 20.6% of the orchid species found in metropolitan France were considered to be supported by "insufficient data" by the IUCN, mainly as a result of problems with taxonomic identification [SCH 14]. Using the integrative taxonomy approach and taking into account, in one analysis, numerous scientifically established taxonomic characteristics (morphology, molecular characteristics, distributional range, flowering period, odor emitted and identity of pollinators), it has progressively become possible, in genus after genus, to clearly identify the taxonomically difficult species. In this context, the variation in the odors emitted by these remarkable flowers is particularly important, because odors of some species attract specific pollinators, whereas those of other species attract a greater diversity of pollinators. These differences have implications for reproductive isolation. Pollinator differences related to variation in floral odor, together with other traits analyzed in the framework of integrative taxonomy, show, for example, how the three described species of the fly orchid group, very close morphologically, can be definitely considered to be distinct species. They differ at the level of their molecular genetics, their habitat preferences, their morphology, the floral scents they emit and their specific pollinators [TRI 13]. Demonstration of these differences shows the importance of implementing conservation programs, because two of these species have restricted distributions (*Ophrys aymoninii*, endemic to the Grand Causses region in France; *O. subinsectifera*, endemic to the Franco-Spanish Pyrenees).

1.2. Scent communication between sexual partners

Another facet of chemical ecology concerns communication between members of the same species. Recent discoveries have shown that certain plants of the African savanna communicate among themselves the arrival of large herbivores; the first plant that suffers their attacks quickly emits a volatile bouquet perceived by neighbors of the same species, which then quickly synthesize protective tannins. In the time it takes an elephant or a giraffe to graze on several leaves, individuals of the same species in the vicinity have already become repulsive [WAR 02].

However, chemical communication within the same species is most developed in animals, especially at the time of reproduction. Reproduction conditions the capacity of species to settle in an environment and colonize it. In animals, scents have a generally determinant role in the recognition, detection (sometimes at long distance) and choice of a sexual partners. To find soul mates, insects have developed amazing olfactory abilities. This is the case with many night-flying moths that are capable of detecting a sexual partner at very long distances. For example, female silkworms (*Bombyx mori*) use bombykol to attract males from within a radius of many kilometers. The males of such species are often equipped with long branching antennae that detect the volatile substances emitted by females [BUT 61]. In *Drosophila* fruit flies, couples find each other on a ripe fruit (Figure 1.2) on which they both come to find food and to use as breeding sites. The mature fruits visited by these flies generally emit phenylacetic acid and phenylacetaldehyde, compounds which act as aphrodisiac stimulants in these small flies. During copulation, the male transmits to the female a pheromone (*cis*-vaccenyl acetate); as a result, future contenders detecting this compound in an already fertilized female can avoid her and optimize their partner selection. Although the production of these perfumes and the molecular mechanisms by which they are detected present differences between vertebrates and invertebrates, how their nervous systems code and decode signals is sometimes strikingly similar. For example, the Asian elephant uses the same sexual pheromone as numerous butterflies [RAS 96]!

Figure 1.2. *A ripe fruit constitutes the ideal place for interaction between male and female drosophila (*Drosophila melanogaster*) flies (left photo: Jean-Pierre Farine). This food source and breeding site is a substrate particularly adapted for finding a sexual partner (right photo: Sonia Dourlot)*

Although the molecules of sexual communication are less well known for vertebrates than for insects, the chemical mediation used by vertebrates includes both small volatile molecules perceived by olfaction and detectable at a distance (such as exo-brevicomin in the mouse) and heavier molecules perceived through contact, such as proteins in the mouse or lipid compounds in several species of Spanish lizards.

A large number of studies have shown that the fragrant bouquets indicate not only the maturity, receptivity and location of the sexual partner but also its quality. Scent variation can even allow olfactory discrimination at the individual scale. It is this property that allows communication of information on the quality of the sexual partner, which is defined, depending on the species, by traits such as its capacity to acquire territory, its social rank, its access to resources, its fertility, its resistance to illness, etc. Numerous examples have been studied not only in cattle and mice (Figure 1.3), but also in social insects (see section 1.5) and primates (see Chapter 3).

Figure 1.3. *Left: scent detection in cattle. Here, a male is sniffing the perigenital zone of a female, to detect volatile compounds linked to female receptivity (photo: F. Urbany). Right: in the domestic mouse (*Mus musculus*), the identity of the sexual partner is perceived via olfaction: a male and a female evaluate each other by sniffing (photo: K. Thonhauser)*

1.3. Scent communication between species

However, it is in the case of communication between species that chemical mediation reaches its richest development, giving free rein to complexity, producing what is sometimes described as chemobiodiversity (defined as the chemical diversity of living things) by analogy to biodiversity. The same scent may sometimes be "interpreted" in different manners by the species involved in the interactions. Immobile and often confronted by the necessity to attract pollinators while at the same time repelling "unwanted" visitors such as herbivores, plants are often the overlooked champions of scent manipulation. In some cases, the same bouquet of floral compounds can simultaneously attract pollinators and repel other visitors. In one such example, male plants of the cycad *Macrozamia* attract their pollinators to cones by the emission of a small quantity of volatile compounds such as β-myrcene [TER 07]. Once the pollinator is attracted, the temperature in the male cones increases, entraining a drastic increase in the emission of β-myrcene and transforming it into a compound that is repulsive to pollinators. These then fly to the cones of female plants, which continue to emit low quantities of β-myrcene. This sequence of visits allows fertilization. Other plants can equally modulate the behavior of insects by using the strategy called "push-pull" (attraction-repulsion), where the plant first emits compounds that attract pollinators, and then (once fertilized) emits compounds repulsive to insects, pollinators as well as herbivores [TER 07]. More generally, floral odor at receptivity, that is, the state in which the flowers are ready to be pollinated, is often different (in quantity emitted and/or in composition) from that emitted before and after pollination.

However, chemical mediators have also evolved to assure that one function may be neutralized, or even subverted, by another organism to serve a quite different function. Organic compounds may be used for defense as well as for communication. Defensive compounds often exert their impact through toxic effects on the target organism(s). Although taste or other properties may provide information about toxicity (for example, many toxic compounds are bitter-tasting), this cannot really be compared with signals in the context of communication. Certain plants assure their defense against herbivores by producing and storing toxic substances such as alkaloids or cyanogenic compounds in their tissues. While these plants are avoided by most insects, certain insects have evolved mechanisms of resistance to these toxins and exploit this "empty ecological niche". Among such insects, some sequester these toxins (e.g. in small vesicles) in their body. The advantage of this sequestration is that it protects these herbivores against their own predators. These herbivores warn of their toxicity by displaying fantastic colors. By evolutionary convergence, this strategy is seen not only in marine environments (e.g. in nudibranchs grazing on algae, coral or sponges) but also in terrestrial environments (e.g. in various butterflies whose caterpillars eat certain plants containing toxic alkaloids). In nudibranchs, this lifestyle presents multiple advantages. These include the following: the lack of need for a shell, allowing greater mobility; widespread use of prey-derived toxins sequestered in diverticula, avoiding the metabolic cost of production or detoxification; extension of the protection from which adults benefit to eggs and larvae; and rapid learning by predators to recognize (often visually) and to avoid these "toxic" prey [ZAG 04]. By convergence, the last three of these advantages also apply in certain terrestrial insects, including Lepidoptera such as Zygaenidae and Nymphalidae and certain beetles (Chrysomelidae). Consequently, the study of chemical mediation between species requires taking into account the environment in which communication takes place and understanding the networks of biotic interactions that are involved.

1.4. Chemical mimicry, to enhance reproduction

The diversity of chemical mediation is based on a large number of organic compounds offering the possibility of an infinite number of combinations between them. These compounds form olfactory bouquets, where the quantity emitted and the relative proportions of these compounds are often pertinent information for the species that perceives this scent. In contrast to sound production, or carrying out a more or less elaborate behavior, the emission of a scent is probably comparatively inexpensive. While scent emission remains an honest signal in numerous circumstances, its low energetic cost leaves the opportunity to use chemical mimicry in ecological situations where the signal transmitted may not be "honest" [BAG 10b].

Chemical mimicry occurs in numerous interactions, and these involve unrelated groups of plant species [HOS 10, BRO 15]. The phenomenon thus bears the hallmarks of convergent evolution. Inflorescences of certain arums, for example, are well known for emitting heat associated with scents of putrefaction or excrement to attract beetles and Diptera (flies), which, in attempting to lay their eggs, are trapped in the scent emission zone. There, they come into contact with the plant's receptive female flowers, which they will pollinate – with pollen that remained on their bodies from their previous visit to another inflorescence of the same species. The pollinators are freed the following day, but the inflorescence morphology is arranged in such a way that on their exit they must pass the male flowers of the inflorescence, where they are dusted with pollen – which they will transport once again to the female flowers of another arum plant. By convergence, certain fungi use a strategy very similar to that of arums, producing a scent of putrefaction that attracts carrion-feeding insects to disperse their spores. Some mosses (notably Splachnaceae and, in particular, *Splachnum luteum*) emit growths strongly resembling corollas, while also producing a fecal scent to attract flies and assure the dispersal of their spores. These arums, fungi and mosses attract their dispersal agents by similar mixtures consisting of mainly monoterpenes (e.g. ocimenes), but also amino compounds (cadaverine and putrescine) and sometimes even sulfurous compounds (di- and trisulfides), with great interspecies variations allowing the capturing of particular "partners" [ALB 08].

Certain species even divert other types of scents; in this domain, orchids display a wide diversity of olfactory deception strategies. There exists first in

Europe the well-known case of nectarless orchids of the genus *Ophrys*, known for the striking visual resemblance of their flowers to certain insects. Flowers of *Ophrys* also emit a scent very similar to the pheromones emitted by females of certain species of Hymenoptera, a scent composed mainly of alkanes and alkenes (non-volatile compounds, detected by antennal contact). The male insects are first deceived visually and then olfactorily, to the point of attempting to copulate with the flowers (pseudocopulation) (Figure 1.4 left). As the insect moves on the flower, the pollinia (pollen packets, typical of orchids) will very often be stuck on the body of the insect and thus transported to another flower, if the insect is deceived again during its search for a sexual partner. In the orchid *Ophrys sphegodes*, the pollinated flower emits a compound, farnesyl hexanoate, which strongly mimics the scent of the female insect after it has been fertilized [SCH 01]. In sexual mimicry, the orchid copies the chemical language of the insect couple; therefore, it is understandable that the male insect confuses the flower and the female of its own species, especially since the orchid generally flowers several days before female insects emerge.

In other Mediterranean orchids, we might at first sight conclude that deception of the pollinator is unnecessary to attract it, since these flowers offer nectar on their labellum (differentiated petal). For example, two new cases of chemical mimicry have recently been demonstrated in species of the genus *Epipactis* (Figure 1.4, right). The broad-leaved helleborine (*Epipactis helleborine*) and the violet helleborine (*E. purpurata*) both emit scents very similar to those emitted when these two species are attacked by an herbivore (aphids) and that correspond to classic plant wound compounds (octanal, hexyl acetate, hexenyl, hexen-1-ol). The advantage for these two species is that this emission attracts the predators of these herbivores, which are wasps. In their active search for prey and then nectar, these wasps will make contact with pollinia, which attach onto the body of the insect and will be transported to another flower. Another species of *Epipactis* (*E. veratrifolia*) extends the trickery even further by imitating the aphid alarm pheromone, emitted during an attack by a predator. The emission of this pheromone – essentially (Z)-11-eicosen-1-ol – attracts hover flies (Diptera) specialized in the capture of these aphids. Once attracted onto the flower, they actively search for their prey and then collect nectar. Pollinia adhere to their bodies and they pollinate other flowers. These orchids thereby divert the predatory behavior of insects to make them effective pollinators.

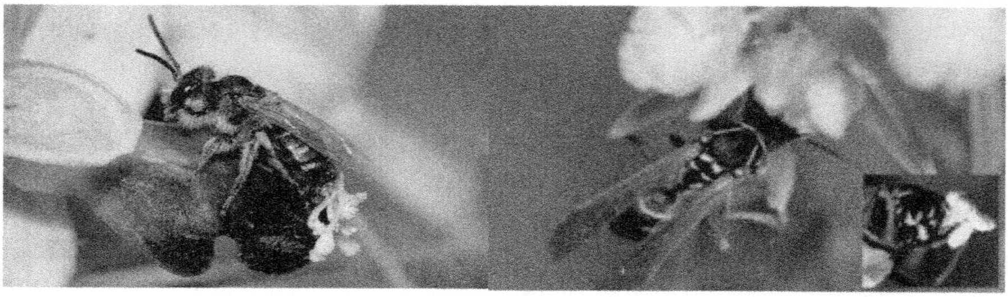

Figure 1.4. *Left: a male bee (*Andrena ovatula*), with the pollinia stuck to the extremity of its abdomen, on the labellum of* Ophrys sulcata*, which emits a scent similar to that of the female of this insect. Right: a wasp (*Polistes nimphus*) on a marsh helleborine flower (*Epipactis palustris*), with a detail of the pollinia adhering to the head of the pollinator insect (photos: Yves Wilcox)*

1.5. A dialog that sometimes evolves into an interaction network

The chemical communication between two species (a transmitter and a receiver) is very often perceived by other species, which can also respond to the initial signal and create a veritable network of chemical interactions. The systems of Russian dolls then organize around a chemical dialog with the phenomena of attraction, manipulation and control. Many "ménages à trois" have been identified and are known to be structured by chemical mediation. For example, this is the case of the mutualist interaction between ants and certain African acacias, where the plant produces swollen hollow structures (domatia) that function as nesting sites, and sometimes also sweet secretions (extrafloral nectar) for feeding the ants, which in turn protect the plant against herbivorous insects. The protective behavior of ants is induced in certain cases by the emission of a scent indicating the moment and the site of attack of the herbivores [SCH 09]. While ant–plant symbioses were long treated as bipartite interactions, recent work has shown that other actors play essential roles (see after). In dry savannas, certain ant species among those inhabiting ant–acacias (e.g. *A. drepanolobium*) emit a strong scent that repels herbivores such as elephants, which devour the leaves of plants lacking ant colonies. When elephants are experimentally excluded (in a fenced reserve), the plant decreases its production of swollen thorns and of food for the ants, and a decrease in protective behaviors of the ants is also observed [PAL 08]. Trees in elephant-excluded enclosures are significantly more attacked by herbivorous insects and suffer greater mortality. This African example shows the dynamic aspect of interspecies mutualism and the role of ants in the maintenance of the landscape [GOH 10]. Many other symbioses between plants and ants exist in humid tropical environments, where often a third mutualist partner intervenes: fungi of the order Chaetothyriales (Ascomycetes) that line part of the inner surface of the domatia. Different lineages of this fungal partner are associated with many different ant–plant interactions throughout the tropics [VOG 11]. One of these is illustrated in Figure 1.5. These fungi use different materials deposited by the plant's specialized and host-specific resident ants (excretions, cadavers, pollen grains, etc.) as food and are then eaten by these ants; they are also involved in the recycling of nitrogen by the plant from the food contributed by the ants [DEF 10]. Recent studies indicate that this inconspicuous fungal partner is crucial to the entire system of interactions.

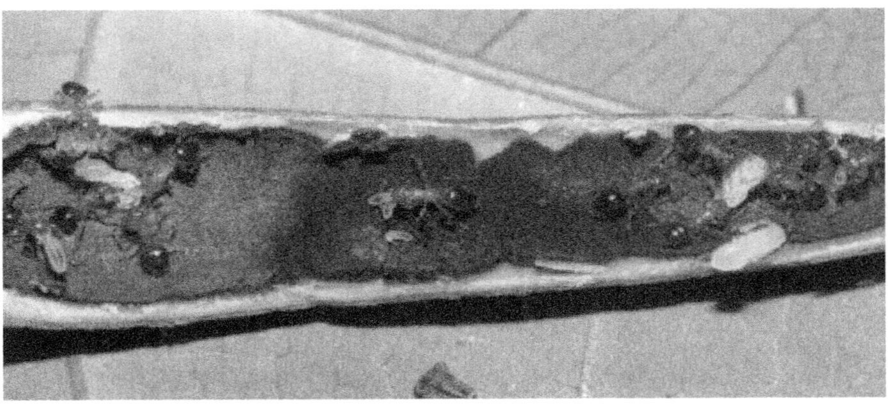

Figure 1.5. *A longitudinal cross-section of domatia, in an internode of the plant* Leonardoxa africana africana, *a small tree in the forests of Cameroon, where the plant–ant* Petalomyrmex phylax *establishes its colonies. A fungus of the order Chaetothyriales is the third partner of this interaction; it is discrete yet visible (the black patch covering part of the inner surface of the domatium) (photo: Rumsaïs Blatrix)*

Another well-known example of symbioses involving three (and sometimes even four) types of organisms is that of lichens (Figure 1.6). Lichens have been viewed as symbiotic associations between heterotrophic fungi and photosynthetic microorganisms, but a third partner, bacteria, has long been ignored. The photosynthetic partner, which assures the energy autonomy of the association, can be a eukaryotic green alga (Viridiplantae) or a cyanobacterium. In the latter case, the photobiont also provides the capacity to fix atmospheric nitrogen. In certain cases, both types of photobionts may co-occur in the same structure (the thallus). Sometimes, these lichens have contrasted morphologies depending on whether the environment favors the expression of the algae or the cyanobacteria. On or in lichen thalli, other microorganisms, particularly (non-photosynthetic) bacteria, are also observed whose number can exceed 10 million individuals per gram of dry lichen. These bacteriobionts can often be visualized owing to their production of characteristic secondary metabolites (depsides, depsidones, aliphatic lactones, etc.), which accumulate on the surfaces of fungal hyphae in the form of crystals, and whose function is so far unknown. The associated organisms appear to have complementary metabolic capacities favoring the general equilibrium of this multi-partner entity: mycobiont/photobiont(s)/bacteriobionts. This symbiosis is characterized by an unequaled capacity to survive in extreme conditions.

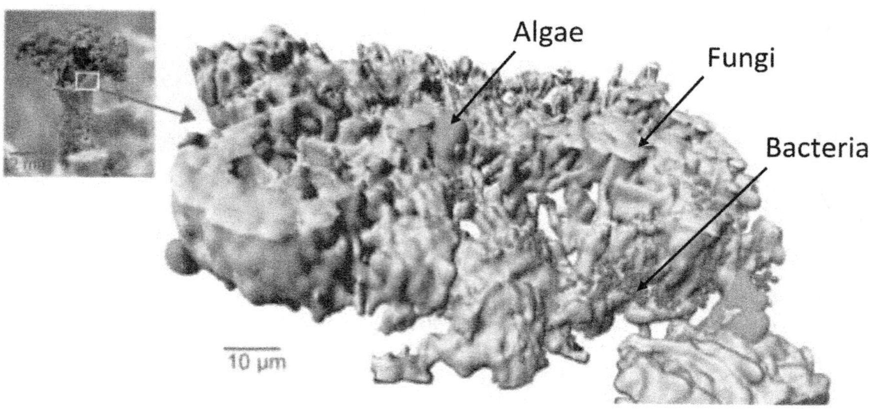

Figure 1.6. *Reconstitution in 3D of layers of confocal scans showing bacterial colonization of the lichen* Cladonia coccifera, *with the fungi, the bacteria and the algae indicated by the arrows (Image: Joël Boustié)*

Other tripartite interactions are not symbiotic, but are temporary associations linked to attacks by herbivores. When infested by the phytophagous mite *Tetranychus urticae*, bean and cucumber plants emit volatile substances that attract the predatory mite *Phytoseiulus persimilis*, which then eliminates *T. urticae*. Mechanisms whereby plants attract enemies of their pests are termed indirect defenses. Discovery of such interactions has not only stimulated fundamental research but has also generated strong interest in the protection of crops using strategies of biological control based on these volatile substances (see Chapter 2). The compounds involved vary among plants: terpenoids in maize and cotton, and sulfur-containing compounds in cabbage and pears. However, the variation is often subtle: for the same plant, the volatile substances emitted differ depending on the species of herbivore that attacked the plant, and even on their stage of development. Jasmonic acid stimulates emission of these volatile compounds and, when it is dusted on a tomato field, stimulates a doubling of the rate of parasitism of a herbivore, caterpillars of the moth *Spodoptera exigua*, by the parasitic wasp *Hyposoter exiguae*. Meanwhile, many tests demonstrate the need to remain prudent about the long-term effects of this strategy, which affects in multiple ways the ecology of each of the species involved in the interaction. For several years, other interactions of this type have also been known in the soil system. For example, phytophagous beetle larvae feeding on maize roots induce the emission by the plant of volatile substances such as β-caryophyllene, which attracts nematode predators of these herbivores! This crop plant is thus at the center of complex interactions involving chemical mediation.

One of the best-studied multi-partner interaction networks is that organized around fig trees. Figs and their pollinators are among the most specialized mutualistic interactions known, in which two species depend on one another for their survival. Other examples are known, including the interactions between certain palm trees and their beetle pollinators, or those between most yuccas and their moth pollinators, but figs offer the most spectacular examples [HOS 10]. With few exceptions, the mutualism between each of the approximately 800 species of figs and the associated pollinating wasp species is obligatory and specific. The plant depends solely on this insect for its pollination and the insect depends on the fig for its reproduction, because it completes its larval cycle in the figs. This specificity is assured by the emission of a scent by the receptive figs (i.e. those in which female flowers are ready to be pollinated) that is detected by the specific pollinator. This scent is often composed of one or more dominant volatile organic compounds, with others present in smaller proportions. The volatile compounds of fig scents are principally mono- and sesquiterpenes, but also include shikimic compounds and fatty acid derivatives. These compounds mediate the encounter between the fig and its specific pollinator, and a variation of this interaction characterizes each of the 800 known species of fig and their pollinating wasps [HOS 10]. However, this mutualism is also exploited by other insects that use the resources exchanged by the mutualists. Among these exploiters are specific parasites of fig wasps and of fig ovules and developing seeds, as well as predatory ants. All of these are capable of detecting scents emitted by figs. Many species of parasites coexist, but each is attracted to figs at a particular stage of development, depending on the resource it exploits. The volatile compounds emitted by figs are crucial for the ability of exploiter insects to recognize the exact stage of fig development at which they can find resources, since figs show little variation in external morphology over development [PRO 07]. Among fig-exploiting insects, those that are parasites are specific to a single species of fig tree, like the pollinator. Depending on the fig species, a dozen or more species of specific parasites are associated with it, the record being 30. For their part, the predatory ants attracted by fig scents capture as prey both fig pollinators and the parasites of figs and pollinators [BAI 13]. The fig tree is, therefore, the support of an interaction network involving numerous specific Hymenoptera, whose coexistence is structured by chemical mediation (Figure 1.7) in a game where the rule is to know who detects the scent of whom. Furthermore, in fig trees, the obligatory and specific pollination mutualism is followed by an opportunistic and less specific mutualism in which fig seeds are dispersed by animals attracted to the plant's fleshy mature fruits. Here again, scents are often important. Mature figs of some species emit scents, usually acetates and alcohol derivatives, which attract numerous seed-dispersing mammals, including bats, terrestrial mammals, monkeys and lemurs. (In figs whose seeds are primarily dispersed by birds, scents appear to be less important in attracting dispersers.) Fig trees play a primordial role in tropical ecosystems, because they produce fruit year-round, and are thus an important food source, particularly in the "lean" season, for many species of birds, bats and primates. Often used by a hundred species of vertebrates, each species of fig tree is the basis of a structured interaction network mediated by organic compounds, enabling figs to play a key role in tropical ecosystems (see also Chapter 2, section 2.2).

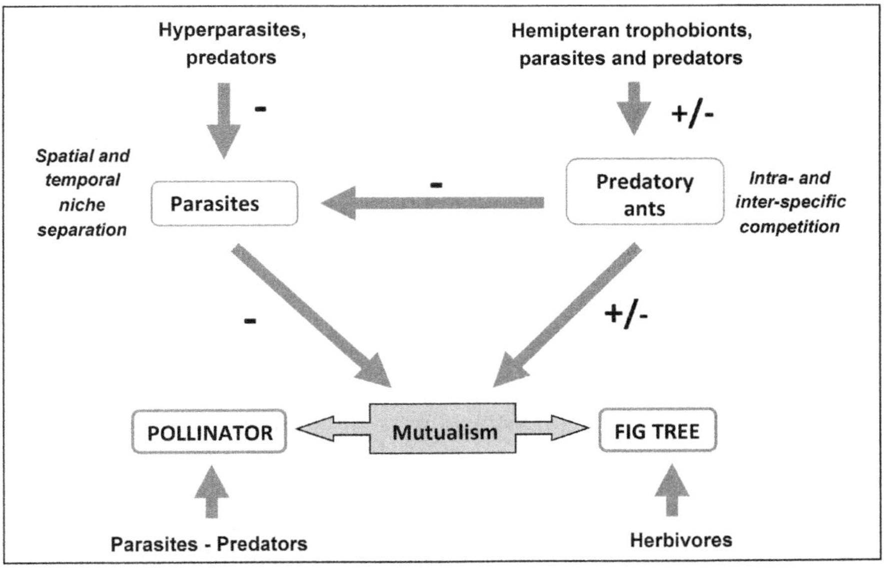

Figure 1.7. *Summary diagram of the range of kinds of organisms associated with fig-pollinator mutualisms. The interactions among the four framed groups of species are structured by chemical mediation*

Other multi-partner interactions, less well known, exist in the microbial world (these are also addressed in Chapter 4). Nutrition is the most frequent benefit at the origin of most interactions between living organisms. The capacity of organisms to use different potential resources is due not only to their own metabolic functions, but also to those supported (or facilitated) by symbiotic microorganisms. Concerning direct defense, it is more appropriate to speak of the use by organisms of informative chemical cues, rather than true communication mediated by chemical signals. By bringing new metabolic capacities coded by genes their hosts do not possess, these microorganisms allow the host to use the particular nutritional resources and to adapt to extreme environments, even to thwart the variations and disturbances in the environment. Taken together, plant-consuming insects use extremely varied nutritional resources, many of which are difficult to break down, toxic, or particularly poor or imbalanced in nutrients. Associated microorganisms allow insects to use such food resources. For example, aphids, by hosting bacteria of the genus *Buchnera* in specialized structures, are able to use plant phloem sap, a resource that is rather deficient in vitamins and essential amino acids. Metagenomic methods have allowed identification of the nutritionally valuable compounds and metabolic precursors (vitamins, fatty acids, amino acids, polyisoprenoids, etc.) that are synthesized by the symbiotic bacteria and supplied to the aphids, and of the simpler resources supplied to bacteria by their host aphid. In parallel with such direct metabolic contributions to their insect hosts, symbiotic bacteria can also influence the physiology of the insect's host plant, enhancing its quality as food [KAI 10]. For example, by influencing the synthesis of phytohormones such as cytokinins, some species of *Wolbachia* bacteria associated with the leaf-mining caterpillar *Phyllonorycter blancardella* can slow down the autumnal nutritional degradation of the leaves of the host plant [KAI 10]. By their anti-senescent action, cytokinins help maintain a favorable nutritional environment, allowing the insect to complete its development even after the leaves have fallen to the ground. This green-island phenomenon is quite widespread, occurring in numerous interactions between plants and organisms that consume them, including insects, fungi and bacteria. Many molecules involved in the response of plants to insects and to microorganisms are hormones, whose exploitation allows certain organisms to take control of the plant for their own needs. Although the molecular mechanisms are usually still unknown, it is clear today that numerous plant phenotypes that had been attributed to effects of insects are in reality induced by the endosymbiotic bacteria of insects, and that chemical mediation plays a privileged role here [LIZ 13].

Such multi-partner interaction networks are also present in aquatic environments, in which organic compounds diffuse in water rather than in air or soil. Molecules functioning in social communication coexist here alongside molecules that mediate competition (biocides or growth inhibitors), preventing the establishment of larvae of other species in the immediate chemosphere of the emitter organism. The emission of molecules of low solubility, such as cyclic terpenoids emitted by soft coral polyps, is often complemented by the secretion of mucus, which facilitates the transport of biocides or growth inhibitors to the neighboring target species. Such interactions appear to be best developed in high-biodiversity habitats such as coral reefs, which are home to a third of all marine biodiversity (Figure 1.8). However, equilibrium of the coral reef ecosystem rests on the good health of the coral builders, which are known to defend themselves or their territory via an arsenal of toxic molecules (terpenes, alkaloids, peptides, etc.). For example, *Tubastraea faulkneri* synthesizes indole alkaloids that inhibit the larvae of other species of coral. Defenses can also act more subtly, for example, by the combination of sterols and fatty acids to prevent implantation of competing larvae or, as in *Heteroxenia fuscescens*, by the emission of indicator molecules in the water to attract symbiotic zooxanthellae. However, the signals emitted by the coral do not pass unperceived by certain predators. For example, the coral species *Porites compressa* emits substances that induce metamorphosis in its predator, the nudibranch *Phestilla sibogae*, enhancing the predator's capacity to exploit its prey. Chemical communication also plays a role in the holobiont, the symbiotic association between coral and zooxanthellae. The coral symbiosis involves mutual recognition and exchange of metabolites between these two partners. A metabolic approach will allow characterizing the physiological state of the holobiont, and finally the identification of compounds involved in the interactions within and between holobionts.

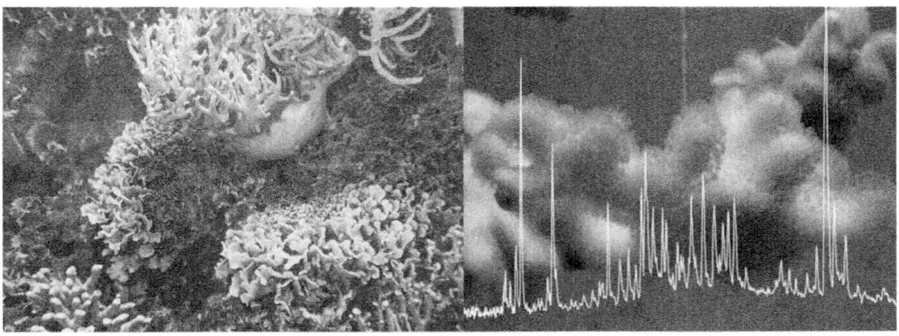

Figure 1.8. *Left:* Sinularia flexibilis *coral (above center) inhibiting the growth of neighboring hard coral Pavona cactus (below) along a gradient of distance between the two (photo: S. La Barre). Right: Longitudinal section of coral (*Stylophora pistillata*) and its characteristic chemical imprint (photo: F. Mohamadi)*

In social animals such as primates (see also Chapter 3) and social insects, intraspecific interactions can also be analyzed as multi-partner interactions. In insects, sociality has produced some of the most remarkable adaptations to the environment known in the entire animal kingdom. The complex societies of social bees and wasps, ants and termites are characterized by three criteria which together define eusociality (true sociality): i) individuals of the same society cooperate in raising young, ii) generations overlap, so that descendants help their parents in raising the young and iii) there is a division of labor, with specialized castes that are morphologically different. In all these aspects, chemical communication within insect colonies is the cornerstone that structures the social system. Ants, termites, and social bees and wasps possess antennae that are truly "olfactory radar", capable of perceiving minute quantities of dozens of compounds, discriminating in these complex mixes of scents the chemical marker elements of the species, the colony, the caste, the sex or even certain physiological states such as fertility.

1.6. Conclusions

The different examples presented in this chapter demonstrate that chemical mediation is crucial for the organization and functioning of biodiversity, terrestrial as well as marine, from microorganisms to the largest of organisms and from species to ecosystems. This mediation currently figures in the global analysis of multidimensional biodiversity, because its study can no longer ignore the new order of multi-partner relations. The bipartite approach to mutualisms has been abandoned in favor of a more global vision, in which the role of each species fits in a continuum between parasitism and mutualism, depending on their respective phonologies and other traits and under the influence of their biotic environment. Both technological and conceptual advances have allowed a better understanding of interaction networks. Additionally, the privileged operation of microbial partners in ecological interactions suggests that the unit on which natural selection acts is not the isolated individual but the holobiont, which is the animal, the plant or both, along with all the associated microorganisms. These complex and fragile interactions are greatly influenced by climatic changes and by increasing pressures exerted by human activities. Against these threats, understanding the complexity of biotic interactions and identifying the organic compounds that mediate them is essential to the conservation of biodiversity.

1.7. Bibliography

[ALB 08] ALBRE J., GIBERNAU M., "Reproductive biology of *Arum italicum* (Araceae) in the south of France", *Botanical Journal of the Linnean Society*, vol. 156, pp. 43–49, 2008.

[BAG 10a] BAGNÈRES A.-G., WICKER-THOMAS C., "Chemical taxonomy with hydrocarbons", in BLOMQUIST G.J., BAGNÈRES A.-G. (eds), *Insect Hydrocarbons: Biology, Biochemistry and Chemical Ecology,* Cambridge University Press, pp. 121–162, 2010.

[BAG 10b] BAGNÈRES A.-G., LORENZI C., "Chemical deception/mimicry using cuticular hydrocarbons", in BLOMQUIST G.J., BAGNÈRES A.-G. (eds), *Insect Hydrocarbons: Biology, Biochemistry and Chemical Ecology*, Cambridge University Press, pp. 282–324, 2010.

[BAI 13] BAIN A., HARRISON R., SCHATZ B., "How to be an ant on figs?", *Acta Oecologica*, vol. 57, pp. 97–108, 2013.

[BOU 14] BOURY-ESNAULT N., LAVROV D.V., RUIZ C.A. *et al.*, "The integrative taxonomic approach applied to *Porifera*: a case study of the *Homoscleromorpha*", *Integrative and Comparative Biology*, vol. 53, pp. 416–427, 2014.

[BRO 15] BROOKER R.M., MUNDAY P.L., CHIVERS D.P. *et al.*, "You are what you eat: diet-induced chemical crypsis in a coral-feeding reef fish", *Proceedings of the Royal Society of London B: Biological Sciences*, vol. 282, 2015.

[BUT 61] BUTENANDT A., BECKAMNN R., HECKER E., "Über den Sexuallockstoff des Seidenspinners. 1. Der biologische Test und die Isolierung des reinen Sexuallockstoffes Bombykol", *Hoppe-Seylers Zeitschriftfür Physiologische Chemie*, vol. 324, p. 71, 1961.

[CAC 15] CACHET N., GENTA-JOUVE G., IVANISEVIC J. *et al.*, "Metabolomic profiling reveals deep chemical divergence between two morphotypes of the zoanthid *Parazoanthus axinellae*", *Scientific Reports*, vol. 5, 2015.

[CAR 12] CARDENAS P., PEREZ T., BOURY-ESNAULT N., "Sponge systematics facing new challenges", *Advances in Marine Biology*, vol. 209, pp. 61–79, 2012.

[DEF 10] DEFOSSEZ E., DJIÉTO-LORDON C., MCKEY D. *et al.*, "Plant-ants feed their host plant, but above all a fungal symbiont to recycle nitrogen", *Proceedings of the Royal Society of London B: Biological Sciences*, vol. 278, pp. 1419–1426, 2010.

[GOH 10] GOHEEN J.R., PALMER T.M., "Defensive plant-ants stabilize megaherbivore-driven landscape change in an African savanna", *Current Biology*, vol. 20, pp. 1–5, 2010.

[HOS 10] HOSSAERT-MCKEY M., SOLER C., SCHATZ B. *et al.*, "Floral scents: their role in nursery pollination mutualism", *Chemoecology*, vol. 20, pp. 75–88, 2010.

[IVA 11a] IVANISEVIC J., THOMAS O.P., LEJEUSNE C. *et al.*, "Metabolic fingerprinting as an indicator of biodiversity: towards understanding inter-specific relationships among *Homoscleromorpha* sponges", *Metabolomics*, vol. 7, pp. 289–304, 2011.

[IVA 11b] IVANISEVIC J., THOMAS O.P., PEDEL L. *et al.*, "Biochemical trade-offs: evidence for ecologically linked secondary metabolism of the sponge *Oscarella balibaloi*", *PLOS One*, vol. 6, no. 11, 2011.

[KAI 10] KAISER W., HUGUET E., CASAS J. *et al.*, "Plant green-island phenotype induced by leaf-miners is mediated by bacterial symbionts", *Proceedings of the Royal Society of London B: Biological Sciences*, vol. 277, pp. 2311–2319, 2010.

[LIZ 13] LIZÉ A., MCKAY R., LEWIS Z., "Gut microbiota and kin recognition", *Trends in Ecology and Evolution*, vol. 28, pp. 325–326, 2013.

[PAL 08] PALMER T.M., STANTON M.L., YOUNG T.P. *et al.*, "Breakdown of an ant–plant mutualism follows the loss of large herbivores from an African savanna", *Science*, vol. 319, no. 5860, pp. 192–195, 2008.

[PIC 06] PICHERSKY E., NOEL J.P., DUDAREVA N., "Biosynthesis of plant volatiles: nature's diversity and ingenuity", *Science*, vol. 311, pp. 808–811, 2006.

[PRO 07] PROFFIT M., SCHATZ B., BORGES R.M. *et al.*, "Chemical mediation and niche partitioning in non-pollinating fig-wasp communities", *Journal of Animal Ecology*, vol. 76, pp. 296–303, 2007.

[RAG 08] RAGUSO, R.A., "Wake up and smell the roses: the ecology and evolution of floral scent", *Annual Review of Ecology, Evolution and Systematics*, vol. 39, pp. 549–569, 2008.

[RAS 96] RASMUSSEN L.E.L., LEE T.D., ROELOFS W.L. *et al.*, "Insect pheromone in elephants", *Nature*, vol. 379, no. 6567, pp. 684, 1996.

[RUI 14] RUIZ C.A., IVANIŠEVIĆ J., CHEVALDONNÉ P. *et al.*, "Integrative taxonomic description of *Plakina kanaky*, a new polychromatic sponge species from New Caledonia (*Porifera*: *Homoscleromorpha*)", *Marine Ecology*, vol. 36, pp. 1129–1143, 2014.

[SCH 01] SCHIESTL F.P., AYASSE M., "Post-pollination emission of a repellent compound in a sexually deceptive orchid: a new mechanism for maximizing reproductive success?", *Oecologia*, vol. 126, pp. 531–534, 2001.

[SCH 09] SCHATZ B., DJIETO-LORDON C., DORMONT L. *et al.*, "A simple, non-specific chemical signal mediates defence behaviour in a specialised ant–plant mutualism", *Current Biology*, vol. 19, pp. 361–362, 2009.

[SCH 14] SCHATZ B., GAUTHIER P., DEBUSSCHE M. *et al.*, "A decision tool for listing species for protection on different geographic scales and administrative levels", *Journal for Nature Conservation*, vol. 22, pp. 75–83, 2014.

[TER 07] TERRY I., WALTER G.H., MOORE C.J. *et al.*, "Odor-mediated push-pull pollination in cycads", *Annual Review of Ecology, Evolution and Systematics*, vol. 39, pp. 549–569, 2007.

[TRI 13] TRIPONEZ Y., ARRIGO N., PELLISSIER L. *et al.*, "Are ecological shifts driving divergence in species of the Fly orchid group?", *Molecular Ecology*, vol. 22, pp. 1431–1446, 2013.

[VOG 11] VOGLMAYR H., MAYER V., MASCHWITZ U. *et al.*, "The diversity of ant-associated black yeasts: insights into a newly discovered world of symbiotic interactions", *Fungal Biology*, vol. 115, pp. 1077–1091, 2011.

[WAR 02] WARD D., YOUNG T.P., "Effects of large mammalian herbivores and ant symbionts on condensed tannins of *Acacia drepanolobium* in Kenya", *Journal of Chemical Ecology*, vol. 28, pp. 921–937, 2002.

[ZAG 04] ZAGROBELNY M., BAK S., RASMUSSEN V.A. *et al.*, "Cyanogenic glucosides and plant–insect interactions", *Phytochemistry*, vol. 65, pp. 293–306, 2004.

2

Chemical Ecology: An Integrative and Experimental Science

All living organisms, whether plants, animals or microorganisms, interact throuch chemical compounds. Chemical ecology seeks to identify the compounds that are involved in these interactions, the structures that allow their biosynthesis, emission and perception, and to decipher their implication in the functioning and the evolution of the living organisms whether at a molecular, individual, population or ecosystemic scale. In addition to fundamental objectives, the understanding of these processes leads to numerous applications in domains as varied as medicine, green chemistry, agriculture or the environment. Concerned with the interactions among organisms at multiple scales, research in chemical ecology relies on a large diversity of experimental approaches.

2.1. Semiochemicals

Semiochemicals are involved in the interactions of organisms among themselves and with their environment. Depending on the type of interaction involved, different terms are used to qualify these mediators that are also called infochemicals or semiochemicals. The term pheromone is used for a semiochemical mediating intraspecific interactions. These pheromones

---

Chapter written by Anne-Marie CORTESERO, Magali PROFFIT, Christophe DUPLAIS and Frédérique VIARD.

which modify the behavior of conspecifics on the short term are called triggering or initiation pheromones. They can also act on the longer term and are then called modifying or induction pheromones. Allelochemicals are involved in interspecific interactions. The word kairomones is used for those which have a positive action on the receiver, allomones is used for those which have a positive action on the emitter and, finally, synomones is used for those which are beneficial to both (Figure 2.1). An entire series of information involved in survival, reproduction, development and use of resources can be exchanged. Understanding this exchange of information often requires the use of experiments in controlled environments in specialized systems adapted to a protagonist lifestyle. *In situ* observations can also be involved. For example, to decipher the chemical language among insects or between insects and plants, devices such as olfactometers or flight tunnels are often used (Figure 2.2). Experiments can also be conducted in greenhouses or directly in the field.

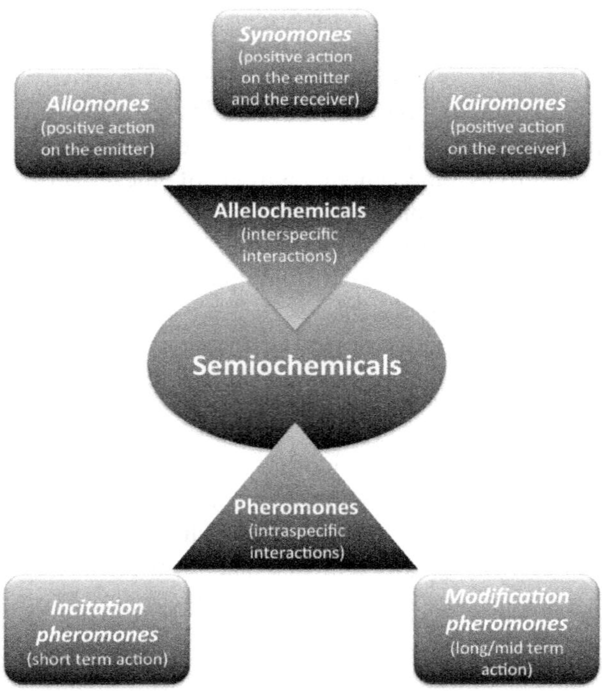

Figure 2.1. *Different types of semiochemicals*

Figure 2.2. *Six-branch olfactometer (photo Thomas Degen)*

On land as in water, organisms communicate via numerous molecules of extremely variable size and complexity. These molecules are, therefore, considered *chemical mediators* (see Figure 2.3). In terrestrial environments, these mediators are often molecules of small size, relatively simple, often lipophilic and volatile, because they are transported by air. Many thousands of molecules have been identified to date; they are generally classified according to the nature of their precursors and their biosynthesis pathways. Four major biosynthesis pathways are generally considered: the acetyl- and malonyl-coenzyme A pathway, at the origin of fatty acids and polyketides, themselves precursors of numerous semiochemicals such as hydrocarbons, hormones or pheromones; the shikimic acid pathway, at the origin of aromatic amino acids and other aromatic compounds like alkaloids; the mevalonic acid pathway, at the origin of isoprenoids and steroids; and finally, the pyruvic acid pathway, at the origin of numerous aromatic compounds such as flavonoids (see also Chapter 5). Molecules involved in defense against bioagressors (like caffeine or strychnine), or protection against UV (such as mycosporine-1) or molt hormones (like ecdysone) act instead by contact, and their structures are often more complex and less hydrophobic. In an aquatic environment, the carrier function of the water favors the transport of a large number of molecules, whether they are soluble or not. Chemical mediators in an aquatic environment are often more complex in terms of size and functions (like didemnin B, produced by a sea squirt, or maitotoxin produced by microalgae, or even halichondrin and vinblastine produced by sponges) (see Chapter 8). The primary or secondary metabolites of organisms living in a marine environment often include elements (like chlorine, bromine, iodine, silicon or boron) from seawater, making them even more complex.

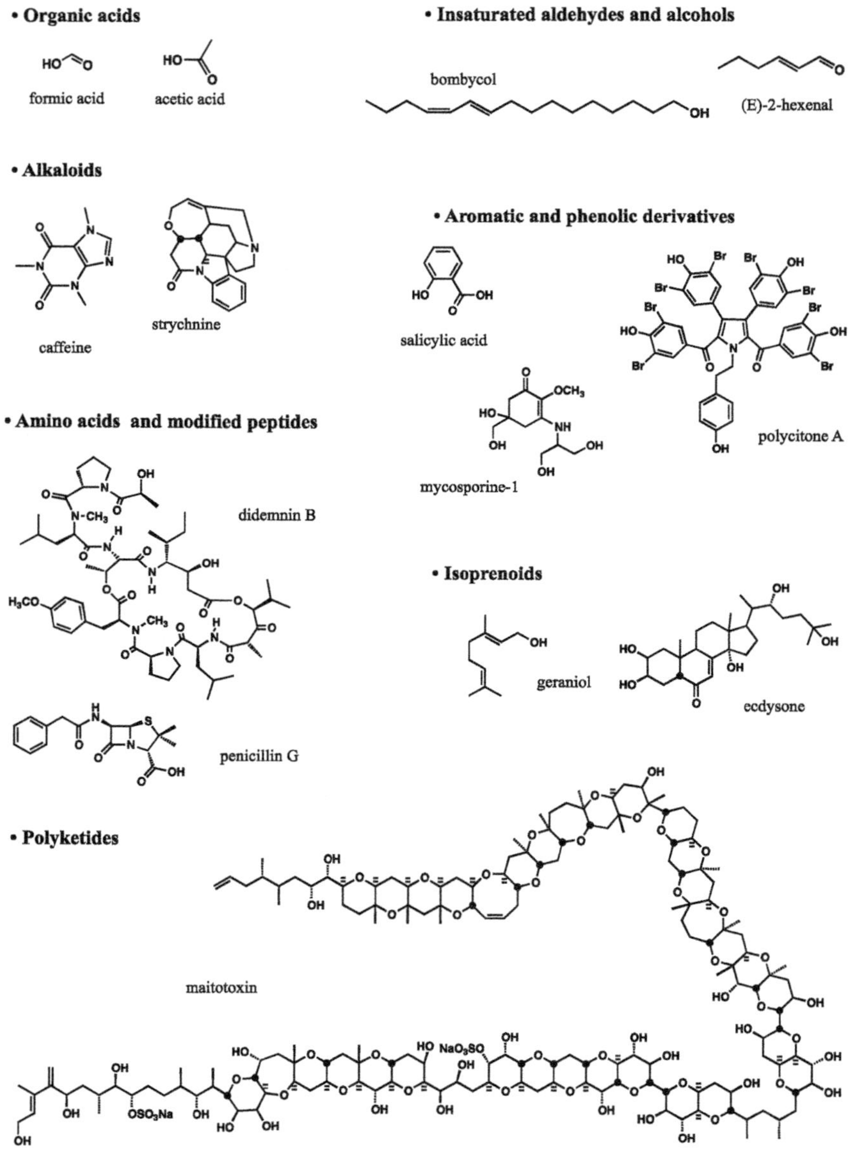

Figure 2.3. *Diversity of semiochemicals (by B. Banaigs)*

The description of these semiochemicals depends on various analytical methods that involve capturing the emitted molecules; separating the constituents from a complex mixture like the liquid–liquid extraction, from the solid phase and using different types of thin layer chromatography, in the liquid phase (HPLC) or in the gaseous phase (GC); and also analyzing the structure of molecules to identify them using mass spectrometry (MS), nuclear magnetic resonance (NMR), infrared (IR) or ultraviolet (UV) spectroscopy. Over the course of a few years, a new technique that allows for dynamic monitoring of compounds emitted even in very small quantities, the PTRMS (Proton Transfer Reaction Mass Spectrometry), has experienced significant development (see Chapter 8).

We estimate today the number of metabolites identified in living organisms at more than 200,000. This diversity reflects the evolution of species and their interactions, in short, biochemical diversity (see Introduction). To understand, inventory and manage this chemical diversity requires the development of complex databases integrating different types of information (like the extract library of ICSN, Cantharella, Ecimar or the eco-chemical library) (see Chapter 8).

Whether in a terrestrial or an aquatic environment, semiochemicals play a determining role in the interactions of a species with its abiotic or biotic environment. The emission and the reception of chemical messengers depend on complex chains of proteins (e.g. enzymes, transport proteins, sensory proteins), whose production and operation can be regulated by local environmental conditions. Indeed, in terrestrial and marine plants, abiotic stresses can have an impact on the emission of chemical messengers through the alteration of the phenology of the plant (growth and maturation), the allocation of resources to the production of these messengers or even via other direct physiological factors such as stomata functioning. If the organs and the mechanisms involved in the reception of the semiochemicals begin to be well understood in animals, the mystery of the perception of these messengers remains yet to be elucidated in plants. The techniques of high-speed analysis (called "omics"), which allow identification of the molecular actors involved (enzymes of biosynthesis pathways, detector proteins, chemical signals) in an exhaustive manner, could contribute to resolving this mystery (see Chapter 5).

For a given intra- or interspecific interaction, the modalities of chemical mediation can vary in time and space. Thus, for example, plants will emit the semiochemical responsible for the attraction of their pollinator(s) only during the period of receptivity of the flowers. After a visit from a pollinator, there is generally a variation in the scented bouquet emitted both quantitatively and qualitatively [HOS 10]. Similarly in animals, the stage of individual development can have an influence on the secondary metabolites emitted. It is the case in certain species of hydrozoans, which use chemical defenses during the larval stage while the adults have a tendency to use structural or physical defenses. These defense compounds are produced by commensal bacteria associated with these hydrozoans, which are more abundant in larvae of these species [HAY 09]. In addition, for receptors of chemical signals, the detection of a particular signal can be modulated in time. This is the case for males of many species of insects, which present an inhibition of their olfactory response to the sexual pheromone of conspecific females during the period following copulation. The modalities of chemical communication can also vary geographically [SOL 11, BON 15]. The intraspecific divergences of chemical signals, like other phenotypical traits, can be the result of two different processes: genetic drift or natural selection allowing adaptation to local conditions. For example, a recent study in a species of orchid demonstrated that adaptation to local pollinators could cause an intraspecific variation of its floral traits, and more specifically of its floral scents [SUN 14]. A geographical variation of food preferences linked to secondary lipophilic metabolites present in certain algae has been demonstrated in a species of polyphagous amphipod herbivore, which uses these metabolites to defend itself against its predators [HAY 09].

2.2. Chemical ecology in multitrophic networks and co-evolution between species

Involved in numerous interactions, semiochemicals play an essential role in the structuring of trophic networks and the evolution of the species. The ecosystems consist of complex networks of multiple interactions. In these networks, the different types of interactions between pairs of species (predation, mutualism or competition) can be specialized or even generalized. The stability of the ecosystems will depend in the first place on the encounter between pairs of species and then in the conservation of these interactions, regardless of the type of interaction. Semiochemicals play a determining role not only at the level of species encounter but also in the maintenance of interactions between species at each level of organization of communities, whether in a terrestrial or marine environment [DIC 10, HAY 09]. Additionally, semiochemicals are involved in the structuring of these communities. For example, in communities of specific and obligatory parasitic wasps associated with fig trees ("non-pollinating fig wasps") (Figure 2.4) (see Chapter 1), it has been demonstrated that semiochemicals play a primordial role in the separation of the trophic niche and, therefore, in the coexistence of these different species on the same resource [HOS 10]. Owing to this, numerous studies in chemical ecology have sought to characterize the compounds involved in these interactions, their role in the organization of multitrophic communities, as well as the evolutionary forces shaping this chemical mediation.

Figure 2.4. *Specific pollinating and parasitic wasps ("non-pollinating fig wasps") on receptive figs of* Ficus racemosa *in India. Semiochemicals play a primordial role in the attraction of these species of wasps to their fig tree host. Predatory ants also use chemical signals to locate their prey on the figs (photo: M. Proffit)*

Technical advances in the area of ecology and chemistry have improved our understanding of the ecological processes mediated by chemical metabolites and, therefore, the communication in the interactions between organisms. From analyses performed in the laboratory, and more precisely using the tools of analytical and synthetic chemistry, many thousands of compounds emitted or stored by organisms, of varying specificity, have been identified to date in terrestrial and marine environments. This large diversity allows a quasi-infinite number of combinations and thus the existence of a multitude of distinct chemical signals unique on the basis of the identity of their compounds. However, for the receiver of the chemical signal, other variables (three-dimensional structure, ratio, and absolute and relative abundance) enter into account when it comes to the recognition of an olfactory signal. In fact, gas chromatography combined with electrophysiological recordings (GC-EAD) has allowed the characterization, among the multitude of compounds emitted by an organism, of the compounds eliciting a response in the receiver of the signal [BRU 05]. This technique, use exclusively in the laboratory, was initially developed in insects. To date, it has been used mainly in the context of the study of chemical communication between organisms in terrestrial arthropods. For example, thanks to electrophysiology, the analytical and synthetic chemistry of the compounds constituting the sexual pheromones emitted by the females of numerous species of moths to attract conspecific males was characterized. Complementing these methods linked purely to the different sensitivities of the chemistry and physiology, the tools of the ecology, mainly of behavioral ecology, allow the characterization of the behavioral responses associated with these metabolites. Currently, a multitude of experimental devices are used in the laboratory (e.g. olfactometer, wind tunnel) as well as in the field (e.g. baited delta traps pitfall traps) for the characterization of the responses of organisms to these semiochemicals.

Although experiments in chemical ecology focus on studying bitrophic or tritrophic interactions, for the past 20 years numerous researchers have attempted to place the results of this work in a multitrophic context (see Chapter 1). For example, numerous works have been investigating the impact of metabolites emitted by plants during an attack by a herbivore (Herbivore-Induced Plant Volatiles, HIPVs) on the other members of the communities surrounding these plants [DIC 10]. Most of these studies were carried out in the laboratory, although a limited number of experiments were conducted in the field. These HIPVs are generally attractive to arthropods, natural enemies of herbivores, but can have a differential impact (attractant or repellent) on the other herbivores associated with these plants, or even on their pollinators. In addition, these HIPVs can interfere with the infestation of plants by pathogens. For example, experiments carried out in a greenhouse showed that transgenic lines of the model organism *Arabidopsis thaliana* not producing certain HIPVs were much more vulnerable to attacks by the fungi *Botrytis cinerea* than lines producing these HIPVs [KIS 08]. After the attack by a herbivore, the locally induced response will propagate in a systematic manner in the entire plant. These defense mechanisms indirectly link the organisms associated with the different parts of the plant. For example, the females of the phytophagous moth *Pieris brassicae* avoid laying their eggs on the leaves of the wild mustard, *Brassica nigra*, when its roots are attacked by the larvae of the cabbage root fly *Delia radicum*, if a non-infested host is present nearby [DIC 10]. In addition, numerous experiments in the laboratory and in the field showed that these HIPVs elicited a defense response in the neighboring plants with which they are con- or heterospecific.

In addition to the purely functional role of these semiochemicals in the organization of the terrestrial and marine communities, much work has concentrated on investigating their adaptive role. In fact, an important challenge for the organisms is to emit and/or interpret these "infochemicals" in an adequate manner to maximize their reproductive success. It is there that chemical ecology requires the tools and concepts of evolutionary ecology. For example, in arthropods, it has been demonstrated that certain compounds secreted by an individual could have multiple functions [BLU 96]. This "semiochemical parsimony" would help to limit the energy costs linked to the biosynthesis of many compounds. Further studies in pollination biology demonstrated that unrelated plants pollinized by the same pollinators presented convergences of floral traits, including the semiochemicals emitted. For example, flowers of plant species pollinated by bats generally emit sulfurous compounds attractive to their pollinators [SCH 13]. Work on nursery pollination mutualisms has underlined the impact of the extreme specificity in these interactions on their semiochemicals. These interactions are mutually beneficial – the pollinator lays its eggs exclusively in one of the structures of its plant host – and generally highly specialized. It seems that the co-evolution between plants and pollinators in these interactions has led to a simplification of the chemical signal produced by plants with a low number of compounds involved in the attraction of the pollinator [HOS 10]. In the domain of plant–herbivore interactions, studies on macro-evolution have also revealed the importance of interactions between species in the evolution of semiochemicals involved in the interaction. For example, the analysis of chemical profiles of 70 species of plants of the genus *Bursera* has demonstrated that, during evolution, the species of this genus would have increased and complexified their chemical defenses [BEC 09]. This complexification of chemical defenses in this genus of plant allowed them to be more resistant to herbivore attacks. Another fascinating example of the impact of co-evolution between plants and herbivores on semiochemicals involved in their interactions is that of butterflies of the Pieridae family associated with plants of different families, with some producing glucosinolates and others not. In Pieridae, a particular detoxification mechanism could have evolved rapidly after the association with plants emitting these glucosinolates and could have contributed to the diversification of this family of butterflies [WHE 07].

2.3. Contribution of chemical ecology to the study of tropical plant diversification

How do we explain the tremendous plant species diversity in Amazonian forests, which are characterized by environmental heterogeneity? First, abiotic factors of each habitat are responsible for the diversification of species in tropical environments and this hypothesis is supported by important beta diversity in Amazonian forests, which was evaluated by comparing the biological diversity along environmental gradients and in different habitats (lowland forests, seasonally flooded forests, white sand forests, inselbergs, savannas, etc.). On the other hand, the diversification of plant species in tropical ecosystems is also correlated with biotic factors. For example, herbivory pressure contributes significantly to plant's specialization into a specific habitat and it has impacts on the natural plant selection. Recently, it has been shown that herbivore insects alter the concurrence between plant species and reduce their spatial distribution while constraining them to develop in sub-habitats only. Herbivore insects, therefore, play a major role in the dynamic of plant adaptation in tropical forests [FIN 04]. In this context, what are the recent contributions of chemical ecology in the study of tropical plant diversification under herbivory pressure and which experiments help reveal the underlying biochemical mechanisms?

The biological interaction between herbivore and plant hosts is an uphill course for chemical, physical and biological armaments. For more than 50 years, chemical ecology has been interested in, among other things, the identification and the evolution of chemical defenses produced by plants (repellent metabolite, insecticide and anti-appetant). This work has sequentially helped lift the veil on the mechanisms of detoxification in insects adapted to defense metabolites. Currently, the chemical phenotypes of living organisms are integrated into global ecological models, which take into account the abiotic and biotic factors. For example, recent work carried out in Peru used reciprocal transplantation gardens to test the "growth-defense trade off" hypothesis in the specialization of tropical plants in a habitat [FIN 06]. The objective of this work is to compare the growth rate, the herbivory rate and the chemical defenses brought into play between communities of specialist species of white sand forests (poor in nutrients) and communities of species having colonized forests rich in nutrients (clay forests). The five-year-old seedlings of 860 seeds belonging to six genera *Mabea* (Euphorbiaceae), *Oxandra* (Annonaceae),

Pachira (Malvaceae *sensu lato*), *Parkia* (Fabaceae), *Protium* (Burseraceae) and *Swartzia* (Fabaceae) have been transplanted in these two habitats. Overall, the concentration of defense metabolites (proteins, phenolic and terpene compounds) does not present any significant variation between the leaves of transplanted plants and those of plants in their original habitats. The chemical defense of species in this study is a fixed trait from a natural selection and its composition does not seem to be affected by resource availability. On the contrary, in habitats poor in nutrients (white sand forests), the transplanted fast-growing species are preferentially attacked by herbivores compared with slow-growing species. The latter allocate the most important part of their resources to chemical defense. This confirms the hypothesis of the defense-growth compromise in the allocation of resources.

A similar study was interested in populations of a single tropical species, *Protium suberratum*, present in various habitats [FIN 13]. Again, the reciprocal transplant gardens were made from populations of white sand forests and populations of terra firma. Unlike the previous study, significant differences were found between the chemical composition of the defense metabolites and their concentration in the leaves. The relative abundance of four classes of chemical products known for their defense role (flavonols, flavones, derivatives of quinic acid and oxidized monoterpenes) varies according to the type of the habitat and the geographic location. Derivatives of quinic acid are present in high concentrations in populations of *P. serratum* of Terra firma (transplanted or non-transplanted), but they are totally absent in populations of white sand forests. Conversely, only the populations of white sand forests possess flavones and oxidized terpenes and these individuals transplanted in Terra firma produce these metabolites in the least proportions. Still, it remains difficult to know whether these astonishing observed differences truly reflect a phenotypic plasticity rather than traits of genetic origin under selection by natural enemies. Nonetheless, these examples illustrate the dynamics of chemical ecology as integrated science in the global studies of evolutionary ecology.

Other notable works focus on volatile organic compounds (VOC) emitted by trees in tropical forests [COU 09]. VOC emitted by the vegetative parts are known for being involved in diverse physiological functions like pollination, and the defense against abiotic and biotic stress; meanwhile, most of the studies have been conducted on model plants, and therefore, the chemical diversity of the VOC at the scale of communities of species remains unknown. Using an SPME (solid-phase micro-extraction) probe, the olfactory bouquet of the leaves and bark was taken from 195 individuals of 55 families present in French Guiana and then analyzed. A total of 264 chemical compounds of distinct biosynthetic origin were detected including 1 compound containing nitrogen, 3 compounds derived from the shikimic acid pathway, 34 compounds from the lipoxygenase pathway, 57 monoterpenes and 169 sesquiterpenes from the terpenoid pathway. The composition of VOC is variable in each family, while the olfactory bouquet is clearly specific to the species, which allowed discrimination of 43 of the 55 studied species by GC–MS analysis (see Chapters 1 and 7). These results reveal, on the one hand, the abundance of sesquiterpenes underestimated previously in the VOC of trees of the Amazon, and, on the other hand, a more important chemical diversity in monoterpenes and sesquiterpenes of VOC emitted by the bark relative to VOC emitted by the leaves [COU 09, COU 12]. This original work is a major contribution to chemical ecology at the scale of communities of species and it offers numerous perspectives in the study of the evolution of chemical defense along an environmental gradient and as a function of herbivory pressure.

Another notable contribution of chemical ecology in the study of tropical plant diversification concerns latex (Figure 2.5). Latex plants represent around 20,000 species of angiosperms (40 families) widely distributed in diverse habitats, with a greater abundance in tropical regions (14%) compared with temperate regions (6%). Latex is an emulsion of polymeric particles (polylactide or polyisoprene) in water and it is found in many parts such as roots, bark, leaves, flowers and fruits. It is also composed of an important chemical diversity of small molecules and proteins, the relative concentrations of which are correlated with biotic and abiotic stress factors. Three principal roles have been attributed to latex: 1) chemical defense (against pathogenic agents, fungi and herbivorous insects), 2) physical defense (injuries and herbivorous insects) and 3) storage (water, metabolic waste and nutrients). The role of latex against herbivorous insects was studied by quantifying the concentration of biomolecules acting as insecticides [KON 11]. Numerous small molecules, such as alkaloids, terpenoids, cardenolides, phenolic compounds and furanocoumarins, as well as numerous insecticide enzymes, such as proteases, oxidases, lectins, chitinases, glucosidases and phosphatases, have been identified. Latex from plants of the Apocynaceae family contains toxic cardenolide metabolites, which can represent up to 30% by mass of dry matter. Cardenolides are inhibitors of Na^+/K^+-ATPases, enzymes playing an important role in the maintenance of electrical potential in most animal cells. They are particularly important in the activity of nerve cells, which explains why cardenolides are toxic for vertebrates and insects. For example, the latex of plants *Antiaris toxicaria* (Moraceae), originating from tropical forests of Southwest Asia, rich in toxicariosides, is used as a poison by the local population for hunting [CAR 97].

Figure 2.5. *Sampling latex in the Amazonian forests. (*Hevea guianensis *and* Chrysophyllum sanguinolentum*) (by C. Duplais)*

Finally, research work on the detoxification mechanism in insects of ouabain, a cardenolide present in latex from the African *Acokanthera ouabaio* (Apocynaceae) plant, has revealed a convergent evolution of adaptation of specialists [HOL 96]. These advances, notable in the chemical classification of latex constituents, are valuable for future investigations into latex phenotypic plasticity and its impact on plant species adaptation in tropical forests.

2.4. When chemical ecology sheds light on the process of biological invasion – an example demonstrating integration between chemistry and ecology

Chemical ecology also sheds light on certain large ecological problems like biological invasions and helps envision new methods for managing invasive species. Biological invasions are one of the pillars of global change. Consequences of deliberate or accidental transport of species by human activities, they modify the composition, biodiversity and functioning of ecosystems worldwide [PER 10b, SIM 09] and are at the center of numerous societal preoccupations [SIM 09]. In parallel with the issues related to conservation of natural environments or of their socio-economic impact, biological introductions are also true natural laboratories for studying a large number of eco-evolutionary processes [SAX 00]. Non-indigenous species are the new players and sometimes disruptors of chemical interactions and mediations already in place in ecosystems. Combined with studies of population ecology, community ecology and experimental ecology, in the field or in the laboratory, chemical ecology sheds major light on these biological invasion processes helping to resolve the paradox of invasion (Figure 2.6) (i.e. How do we explain the success of a species in an environment new to it?).

The chemical signals emitted in the environment by local species play a role from the beginning of the introduction process, notably for the detection of habitats favorable to the establishment of non-indigenous species. For example, in *Crepidula fornicata*, a marine gastropod of North American origin and invasive in Europe (Figure 2.7), bioassays have shown that dibromo-methane, emitted by certain red algae, act as an inductor of the metamorphosis of larvae of this invasive species [TAR 10]. Under the effect of these molecules, particular metabolic pathways such as the "*nitric oxide synthase*" are activated during larval development and metamorphosis. These halogen compounds could be markers of favorable habitats for numerous species of invertebrates and play an important role in the colonization dynamic of new habitats. Semiochemicals could even favor the spread of introduced species introductions, notably from the fact of very specific predator-prey associations. This hypothesis was proposed for the *Syphonota geographica* aplysia introduced via the Suez Canal (lessepsian migration). This mollusc displays compounds identical to those of the marine phanerogam *Halophila stipulacea*, introduced previously in the same geographic area [MOL 08]. The mollusc is a specialist herbivore of this marine plant: this specialization could have "facilitated" the expansion of the mollusc distribution area by following the chemical path of the marine plant.

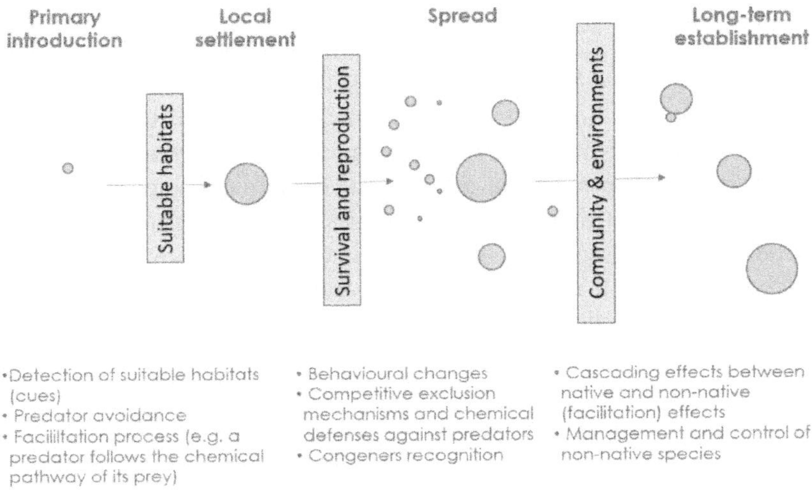

Figure 2.6. *Schematization of the different stages (top labels) and filters (text in boxes) operating during a biological invasion. At each of these stages and to cross these barriers and filters, a diversity of signals and chemical cues will be brought into play. The blue text illustrates by several examples the processes in which these chemical interactions can play a key role*

Figure 2.7. *Left: Chemical signals will allow this small swimming mollusc larva (400–600 µm)* Crepidula fornicata *to find its established congeners at the bottom of the sea after a journey which would last for several weeks depending on the currents (photo: Robin Stolba, Yann Fontana – Station Biologique de Roscoff). Right:* Bugula neritina *is a small animal of the group of bryozoans, which lives by forming colonies resembling small-branched arbuscules. Largely transported by human activities, this species today presents a cosmopolitan distribution and has been the object of numerous studies in chemical ecology (e.g. role of bryostatin as a repellent against predators) (photo: Wilfried Thomas – Station Biologique de Roscoff)*

Another very important aspect in which chemical ecology aided significant advances is the study of behavioral changes observed in numerous non-indigenous species following their introduction. The most demonstrative cases include colonies of social insects, particularly the evolution of unicolonial populations, called, in certain cases, super colonies. For example the Argentine ant (*Linepithema humile*), the small fire ant (*Wasmannia auropunctata*) and the fire ant (*Solenopsis invicta*), and also the underground termites of Europe of invasive origin such as *Reticulitermes urbis* and *R. flavipes*. A relationship has been established for numerous species of a very polygynous nature and marked by low aggressivity of the colonies, and the homogeneity of recognition of individuals, of long-chain cuticular hydrocarbons forming a chemical signature, to the difference of non-invasive species and native populations [PER 10a, PER 11] (Figure 2.8).

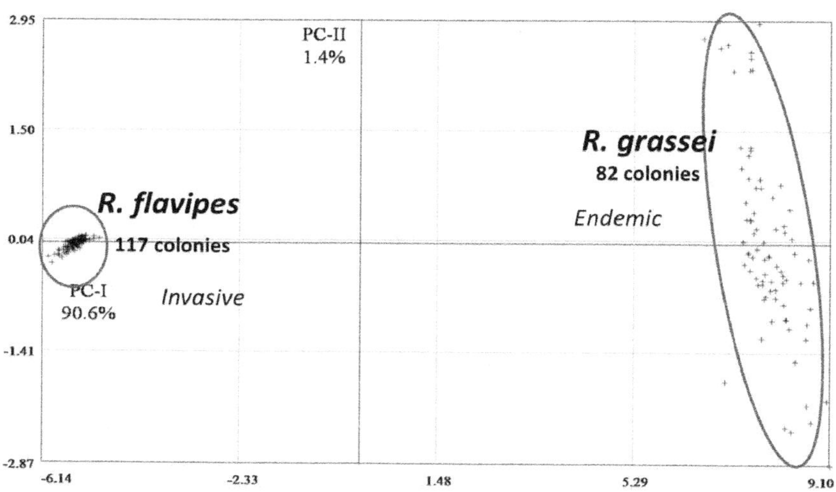

Figure 2.8. *Comparison of chemical signatures of two termite species present in Europe, an invasive,* Reticulitermes flavipes, *originating from the USA, and an endemic,* R. grassei *(from [PER 11])*

The sustainable establishment of non-indigenous species also involves the establishment of new relationships, notably competitive or defensive, with the species already present in the environment. It is, therefore, not surprising that studies on the allelopathic activity, in particular in terrestrial plants, have been particularly numerous [IND 11].

These studies led to the proposal of a major hypothesis in the domain of invasive biology: the "novel weapons hypothesis" (NWH). This hypothesis predicts that the success of introduced species should result in the presence of unique secondary metabolites in this species in their area of introduction. This hypothesis was notably proposed to explain the processes of competitive exclusion by the "diffuse knapweed" *Centaurea diffusa*, of which the root exudates are phytotoxins for numerous other indigenous plants. This type of interaction had also been demonstrated in other environments, notably marine. For example, the red algae *Bonnemaisonia hamifera*, particularly invasive in Scandinavia, produces a unique secondary compound, called 1,1,3,3-tetrabromo-2-heptatone, which inhibits the recruitment of spores and propagules of indigenous algae, reducing competition with local species [ENG 09]. In aquatic animals, these mechanisms of competitive exclusion can also be exerted through strategies of avoidance and escape, as has been demonstrated in South Africa by the study of the trajectories of different indigenous gastropods in response to the presence of water preconditioned with the non-indigenous gasteropod *Tarebia granifera*. These semiochemicals are also involved in defense against predators. It is in this manner that 1,1,3,3-tetrabromo-2-heptatone emitted by *B. hamifera* is a repellent to predators. Also, the larvae and adults of the invasive species *Bugula neritina* are protected from predation by chemical defenses linked to the production of a macro-molecule called bryostatin (macrocyclic lactone), which is produced by endosymbiotic gamma proteobacteria (Figure 2.7).

Semiochemicals can also have cascading effects, as in the case of the Amur honeysuckle, a tree native to Asia and invasive to the United States, whose roots emit phenolic compounds and whose leaves inhibit the growth of indigenous plants, entraining a modification of the chemical composition of the litters, and finally the nearby ponds. This ultimately induces the accelerated development of the tadpoles of the American toad *Anaxyrus americanus*. The effects of these chemical signals can also be very specific. Thus, in response to these injuries, the invasive green algae *Codium fragile* (ssp. *tomentosoides*) emits dimethyl sulfone propionates which, with their products of degradation (dimethyl sulfide and acrylic acid), have a repellent effect on the sea urchin *Strongylocentrotus droebachiensis*, protecting it against attacks by this herbivore. However, this defense does not work with the herbivore mollusc *Littorina littorea*. In the rocky basins of Nova Scotia, the abundance of this mollusc, therefore helps limit the proliferation of the alga *C. fragile* (ssp. *tomentosoides*). The studies of chemical ecology, and more particularly, the metabolic approaches, could eventually provide the tools to control the populations of invasive species (see Chapter 7).

The use of semiochemicals for intraspecies communication, such as pheromones, represents a path of promising study to fight against the proliferation of some of these invasive species. A study has been conducted in Australia on the tadpoles of the *Bufo marinus* toad, a species particularly invasive on this continent. The tadpoles emit alarm pheromones, called bufodienolides, toward one another, which have no effect on other species of local frogs [HAG 09]. This specific pheromone could be used as a tool for biochemical control because it reduces the survival of tadpoles of the *B. marinus* toad as well as their size at metamorphosis increasing their vulnerability to predators and parasites. Another way of controlling them consists of using existing cannibalism in this species [CRO 11]: the tadpoles consume the eggs of their own species by identifying them through specific chemical signals that they emit. In sea lampreys, an anadromous species (i.e. a species born in freshwater, then spending most of its life in the sea and finally returning to freshwater to spawn), steroidal compounds play the role of migratory pheromones. Their effectiveness at a very low dose has been demonstrated along with the use of these proposed compounds to reduce the size of populations on nesting sites in the Canadian Great Lakes region, where the species was introduced around a century ago with effects on local fisheries.

These different examples illustrate the diversity of applications of chemical ecology to the biology of invasions, from the understanding of the ecological and evolutionary processes involved in the success of the introduction of non-indigenous species to the proposal of "eco-biochemical" control solutions. The biology of invasions is also a research domain illustrating the importance of the integration between experimental ecology, field ecology and chemical ecology.

2.5. Protection is in the air: how plants defend themselves against phytophagous insects through VOC emissions

Beyond understanding plants/pest insect interactions, chemical ecology can help design more environmentally friendly crop protection methods. The VOCs emitted by plants are key mediators in the search and use of resources by predators and parasitoids (see Chapter 9, particularly Figure 9.4). Indeed, these compounds (which can be emitted at the level of the attacked parts of the plant or systemically by the whole plant) can help locate and identify their phytophagous hosts or prey at a distance [DIC 10]. Thus, the attacks on tobacco plants by two close species of caterpillars, *Heliothis virescens* and *Helicoverpa zea*, induce the emission of a bouquet of VOCs composed of the same molecules, but at different concentrations. This is a difference that the parasitoid *Cardiochiles nigriceps*, which only oviposits in *Heliothis virescens*, perceives. Such VOCs therefore play an essential role in the recruitment of these natural enemies but their influence does not stop there, since they also influence their oviposition decisions (and, therefore, host infestation) or their time of residence in the patch. These volatile emissions from plants are generally considered part of their indirect defenses [TUR 04]. However, these VOCs can also be used by phytophagous insects to optimize their own use of resources [BRU 05]. They play a major role in locating host plants from a distance in numerous herbivores, as for example, in the apple maggot *Rhagoletis pomonella* or the onion maggot *Delia antiqua* in which the compounds involved have been precisely identified [MIL 90]. They also influence the acceptance of this plant and stimulate oviposition. These are essential signals in decision-making by such insects. This is the case in particular for HIPVs, which we have mentioned already. Indeed, as they are emitted after an initial herbivore attack, they can give information on the quality of oviposition or feeding sites, indicating the presence of competitors or natural enemies. They can, therefore, repel herbivores and form part of the plants' direct defense.

At a time when the use of phytosanitary products against pest insects shows numerous limitations, the major influence of VOCs emitted by plants on the behavior of phytophagous insects and their natural enemies helps envision their use in the development of new strategies of crop protection more suitable for the environment and human health [TUR 06] (see Chapter 9). These compounds can be used in the form of synthetic compounds placed directly in dispensers to protect crops. This is the case, for example, for methyl salicylate used in various crops such as hop yards, vineyards, sweet corn, broccoli, wine grapes and cotton to attract parasitoids and predators and reduce the populations of important pests such as aphids and phytophagous mites [ROD 11]. Some of these compounds can simultaneously influence an insect pest and its natural enemies. This is the case with DMDS, a sulfurous compound emitted in particular by Brassicaceae infested by the cabbage maggot, *Delia radicum*. Placed in dispensers in experimental plots of broccoli, this compound greatly reduces egg-laying of the insect pest and attracts its natural enemies, the rove beetles *Aleochara bilineata* and *A. bipustulata* [FER 09] (Figure 2.9). Another strategy consists of directly using the plants which emit attractive or repellent compounds as companion plants or in push–pull (or "stimulo deterrent diversion") strategies [COO 06] (see Chapter 1). This latter strategy was implemented in East Africa to protect maize fields against several species of stem borer Lepidoptera (*Chilo partellus*, *Eldana saccharina*, *Busseola fusca* and *Sesamia calamistis*). In this strategy, a plant emitting repellent compounds (*Desmodium uncinatum*) is placed in the crop to protect, while another plant (*Pennisetum purpureum*) emitting attractive compounds is placed in a border belt to act as trap plant for the pest insects. An intermediate strategy consists of triggering the emission of VOCs of interest (acting on the pest insects or on their natural enemies) by applying compounds called elicitors to the plant [HOW 08]. These can be phytohormones such as jasmonic acid, which is part of the cascade of reactions induced by the attack of an herbivore on a plant. An application of this compound on healthy plants induces the emission of signals attractive for the herbivores' natural enemies in plants such as bean (*Phaseolus lunatus*), gerbera (*Gerbera jamesonii*), tomato (*Lycopersicon esculentum*), cotton (*G. hirsutum*) and thale cress (*Arabidopsis thaliana*).

Figure 2.9. *Use of synthetic VOCs to manipulate the behavior of a pest insect (the cabbage maggot,* Delia radicum*) and its natural enemies (*Trybliographa rapae *and* Aleochara bilineata*) and protect brassicaceous crops (photos: Sonia Dourlot and Bernard Chaubet – UMR IGEPP, Rennes)*

If the applied potential of these semiochemicals is enormous, actual applications remain rare (see Chapter 9) and require more research, both to better target the mechanisms to the work, and to adapt these innovative strategies to agriculture practices and support their adoption.

2.6. Conclusions

Dealing with the study of chemical interactions in the living world at different scales, chemical ecology is necessarily integrative. It feeds the permanent back and forth between experiments in the laboratory and observations in the field. This booming scientific domain benefits from several years of major technological advances in the domains of life sciences (e.g. the so-called "-omics" technologies) and chemistry (e.g. PTRMS and high resolution RMN). These advances have brought increased accuracy to the identification of mediators and chemical signals and improved their methodological integration. They have contributed to significant advances in the understanding of interactions within a single species or among multiple species and, therefore, in our capacity to decode certain co-evolutionary mechanisms acting on the structure and dynamics of populations and communities of marine and terrestrial organisms. Research in chemical ecology also helps respond to actual problems resulting from global changes and contributes to sustainable development.

2.7. Bibliography

[BEC 09] BECERRA J.X., NOGE K., VENABLE D.L., "Macroevolutionary chemical escalation in an ancient plant-herbivore arms race", *Proceedings of the National Academy of Sciences*, vol. 106, pp. 18062–18066, 2009.

[BLU 96] BLUM M.S., "Semiochemical parsimony in the Arthropoda", *Annual Review of Entomology*, vol. 41, pp. 353–374, 1996.

[BON 15] BONELLI M., LORENZI M.C., CHRISTIDÈS J.-P. *et al.*, "Population diversity in cuticular hydrocarbons and mtDNA in a mountain social wasp", *Journal of Chemical Ecology*, vol. 41, pp. 22–31, 2015.

[BRU 05] BRUCE T.J., WADHAMS L.J., WOODCOCK C.M., "Insect host location: a volatile situation", *Trends in Plant Science*, vol. 10, pp. 269–274, 2005.

[CAR 97] CARTER C.A., FORNEY R.W., GRAY E.A. *et al.*, "Toxicarioside A. A new cardenolide isolated from *Antiaris toxicaria* latex-derived dart poison. Assignment of the 1H- and 13C-NMR shifts for an antiarigenin aglycone", *Tetrahedron*, vol. 53, pp. 13557–13566, 1997.

[COO 06] COOK S.M., KHAN Z.R., PICKETT J.A., "The use of push-pull strategies in integrated pest management", *Annual Review of Entomology*, vol. 52, p. 375, 2006.

[COU 09] COURTOIS E.A., PAINE T.C.E., BLANDINIERES P.-A. *et al.*, "Diversity of the volatile organic compounds emitted by 55 species of tropical trees: a survey in French Guiana", *Journal of Chemical Ecology*, vol. 35, pp. 1349–1362, 2009.

[COU 12] COURTOIS E. A., BARALOTO C., PAINE T.C.E. *et al.*, "Differences in volatile terpene composition between the bark and leaves of tropical tree species", *Phytochemistry*, vol. 82, pp. 81–88, 2012.

[CRO 11] CROSSLAND M.R., SHINE R., "Cues for cannibalism: cane toad tadpoles use chemical signals to locate and consume conspecific eggs", *Oikos*, vol. 120, pp. 327–332, 2011.

[DIC 10] DICKE M., BALDWIN I.T., "The evolutionary context for herbivore-induced plant volatiles: beyond the 'cry for help'", *Trends in Plant Science*, vol. 15, pp. 167–175, 2010.

[ENG 09] ENGE S., NYLUND G.M., HARDER T. *et al.*, "A novel chemical releases the invasive red alga *Bonnemaisonia hamifera* from new enemies", *Phycologia*, vol. 48, p. 29, 2009.

[FER 09] FERRY A., LE TRON S., DUGRAVOT S. *et al.*, "Field evaluation of the combined deterrent and attractive effects of dimethyl disulfide on *Delia radicum* and its natural enemies", *Biological Control*, vol. 49, pp. 219–226, 2009.

[FIN 04] FINE P.V., MESONES I., COLEY P.D., "Herbivores promote habitat specialization by trees in Amazonian forests", *Science*, vol. 305, pp. 663–665, 2004.

[FIN 06] FINE P.V., MILLER Z.J., MESONES I. *et al.*, "The growth-defense trade-off and habitat specialization by plants in Amazonian forests", *Ecology*, vol. 87, pp. S150–S162, 2006.

[FIN 13] FINE P.V., METZ M.R., LOKVAM J. *et al.*, "Insect herbivores, chemical innovation, and the evolution of habitat specialization in Amazonian trees", *Ecology*, vol. 94, pp. 1764–1775, 2013.

[HAG 09] HAGMAN M., SHINE R., "Larval alarm pheromones as a potential control for invasive cane toads (*Bufo marinus*) in tropical Australia", *Chemoecology*, vol. 19, pp. 211–217, 2009.

[HAY 09] HAY M.E., "Marine chemical ecology: chemical signals and cues structure marine populations, communities, and ecosystems", *Annual Review of Marine Science*, vol. 1, pp. 193–212, 2009.

[HOL 96] HOLZINGER F., WINK M., "Mediation of cardiac glycoside insensitivity in the monarch butterfly (*Danaus plexippus*): role of an amino acid substitution in the ouabain binding site of Na^+, K^+-ATPase", *Journal of Chemical Ecology*, vol. 22, pp. 1921–1937, 1996.

[HOS 10] HOSSAERT-MCKEY M., SOLER C., SCHATZ B. *et al.*, "Floral scents: their roles in nursery pollination mutualisms", *Chemoecology*, vol. 20, pp. 75–88, 2010.

[HOW 08] HOWE G.A., JANDER G., "Plant immunity to insect herbivores", *Annual Review of Plant Biology*, vol. 59, pp. 41–66, 2008.

[IND 11] INDERJIT W., WARDLE D.A., KARBAN R. *et al.*, "The ecosystem and evolutionary contexts of allelopathy", *Trends in Ecology and Evolution*, vol. 26, pp. 655–662, 2011.

[KIS 08] KISHIMOTO K., MATSUI K., OZAWA R. *et al.*, "Direct fungicidal activities of C6-aldehydes are important constituents for defense responses in Arabidopsis against *Botrytis cinerea*", *Phytochemistry*, vol. 69, pp. 2127–2132, 2008.

[KON 11] KONNO K., "Plant latex and other exudates as plant defense systems: roles of various defense chemicals and proteins contained therein", *Phytochemistry*, vol. 72, pp. 1510–1530, 2011.

[MIL 90] MILLER J.R., COWLES R.S., "Stimulo-deterrent diversion: a concept and its possible application to onion maggot control", *Journal of Chemical Ecology*, vol. 16, pp. 3197–3212, 1990.

[MOL 08] MOLLO E., GAVAGNIN M., CARBONE M. *et al.*, "Factors promoting marine invasions: a chemoecological approach", *Proceedings of the National Academy of Sciences USA*, vol. 105, pp. 4582–4586, 2008.

[PER 10a] PERDEREAU E., DEDEINE F., CHRISTIDÈS J.-P. *et al.*, "Variations in worker cuticular hydrocarbons and soldier isoprenoid defensive secretions within and among introduced and native populations of the subterranean termite, *Reticulitermes flavipes*", *Journal of Chemical Ecology*, vol. 36, pp. 1189–1198, 2010.

[PER 10b] PERRINGS C., MOONEY H., WILLIAMSON M., *Bioinvasions and Globalization – Ecology, Economics, Management and Policy*, Oxford University Press, Oxford, 2010.

[PER 11] PERDEREAU E., DEDEINE F., CHRISTIDÈS J.-P. *et al.*, "Competition between invasive and indigenous species: an insular case study of subterranean termites", *Biological Invasions*, vol. 13, pp. 1457–1470, 2011.

[ROD 11] RODRIGUEZ-SAONA C., KAPLAN I., BRAASCH J. *et al.*, "Field responses of predaceous arthropods to methyl salicylate: a meta-analysis and case study in cranberries", *Biological Control*, vol. 59, pp. 294–303, 2011.

[SAX 00] SAX D.F., BROWN J.H., "The paradox of invasion", *Global Ecology and Biogeography*, vol. 9, pp. 363–371, 2000.

[SCH 13] SCHIESTL F.P., JOHNSON S.D., "Pollinator-mediated evolution of floral signals", *Trends in Ecology & Evolution*, vol. 28, pp. 307–315, 2013.

[SIM 09] SIMBERLOFF D., "We can eliminate invasions or live with them. Successful management projects", *Biological Invasions*, vol. 11, pp. 149–157, 2009.

[SOL 11] SOLER C., HOSSAERT-MCKEY M., BUATOIS B. *et al.*, "Geographic variation of floral scent in a highly specialized pollination mutualism", *Phytochemistry*, vol. 72, pp. 74–81, 2011.

[SUN 14] SUN M., GROSS K., SCHIESTL F.P., "Floral adaptation to local pollinator guilds in a terrestrial orchid", *Annals of Botany*, vol. 113, pp. 289–300, 2014.

[TAR 10] TARIS N., COMTET T., STOLBA R. *et al.*, "Experimental induction of larval metamorphosis by a naturally-produced halogenated compound (dibromomethane) in the invasive mollusc *Crepidula fornicata* (L.)", *Journal of Experimental Marine Biology and Ecology*, vol. 393, pp. 71–77, 2010.

[TUR 04] TURLINGS T.C.J., WÄCKERS F., "Recruitment of predators and parasitoids by herbivore-injured plants", in CARDÉ R.T., MILLAR J.G. (ed.), *Advances in Insects Chemical Ecology*, Cambridge University Press, Cambridge, UK, pp. 21–75, 2004.

[TUR 06] TURLINGS T.C.J., TON J., "Exploiting scents of distress: the prospect of manipulating herbivore-induced plant odours to enhance the control of agricultural pests", *Current Opinion in Plant Biology*, vol. 9, pp. 421–427, 2006.

[WHE 07] WHEAT C.W., VOGEL H., WITTSTOCK U. *et al.*, "The genetic basis of a plant-insect coevolutionary key innovation", *Proceedings of the National Academy of Sciences*, vol. 104, pp. 20427–20431, 2007.

3

Scents in the Social Life of Non-Human and Human Primates

One major evolutionary transition in the history of life is the aggregation of solitary individuals into structured societies. The organization and maintenance of these animal societies are possible thanks to elaborate mechanisms of interindividual communication. Chemical communication is a widespread mechanism used by most living species, including human and non-human primates, incorrectly considered as "microsmatic" species. In these primates, body odors may reflect individual characteristics such as sex, dominance rank, genetic quality or health. In addition, the perception of allelochemical substances (involved in plant defenses) allowed feeding habits to evolve, each species carefully choosing, or selectively avoiding, food items that fit their physiological needs. Such processes of selection/avoidance also allowed the progressive emergence of therapeutic practices, particularly in humans. Therefore, chemical signals should be regarded as key elements in the organization of most societies, from insects to primates, including humans.

3.1. Primate societies and their complex systems of communication

One of the major evolutionary transitions in the history of life is the aggregation of solitary individuals into organized and structured societies. Social systems are, indeed, emerging properties of individual behavioral strategies. The structure and maintenance of these systems are possible thanks

---

Chapter written by Marie CHARPENTIER, Guillaume ODONNE and Benoist SCHAAL.

to elaborate processes of interindividual communication. As such, the "social complexity" hypothesis stipulates that group life has influenced the evolution of communication systems toward a complexification of signals [FRE 12]. For example, a comparative analysis of 22 species of Sciuridae (e.g. prairie dogs) showed that the degree of social complexity was positively correlated with the number of different alarm calls from the acoustic repertoire. In 42 species of non-human primates, the size of the vocal repertoire also positively co-varied with several estimates of social complexity. In 12 other primates, group size was positively correlated to the diversity of facial expressions [FRE 12]. While the relationship between social complexity and the complexity of communication systems has been studied in numerous taxa for the visual and acoustic channels, this is not the case for chemical signals. The few exceptions include lemur's olfaction. In these strepsirrhine primates, comparative analyses of the chemistry of glandular secretions and urines showed that the characteristics of the social organization and of the dominance hierarchy were both linked to the richness in specific odorants (e.g. [DEL 12]). In this chapter, we aim to show that chemical communication, including olfaction and gustation, plays a central role in structuring primate social relationships, including "microsmatic" Old World species including humans.

The primate order is characterized by an extreme diversity of social systems. Some aspects of this diversity include patterns of individual aggregation and modes of reproduction, and also an extreme diversity and complexity of social relationships. The characteristics of these social interactions, such as their quality, intensity, direction or durability, are determined by several factors including, for example, dominance relationships, genetic relatedness and/or mating patterns. Living in societies has entailed a diversification of behavioral strategies requiring a complexification of communication systems. Primates have, therefore, developed numerous mechanisms throughout evolutionary time, including acoustic, visual, gestural and chemical communication.

Chemical communication is a ubiquitous system that may occur through body contacts or close interactions, and also at distance by scent-marking of substrates. Thus, the absence of the emitters does not break the communication with the receivers of these scents. Chemical communication is also a language that can be decoded in the absence of light and, as such, constitutes a prime mode of communication in nocturnal species. In primates, this system of communication has long been ignored but has recently received a renewed interest in "macrosmatic" species characterized by a well-developed sense of smell. Among them, lemurs display an extremely sophisticated chemical communication system, where socio-sexual relationships between individuals depend greatly on the information encoded in the chemical signals emitted by the two sexes. For example, in ring-tailed lemurs (*Lemur catta*), males possess three glands located on the scrotum, wrists and armpits. Females possess a single labial gland. These four glands produce distinct odors: secretions from the labial gland and the scrotum contain mostly organic acids and esters while secretions of male's wrist glands are mainly composed of terpenoids and steroids (squalene, cholesterol derivatives, lanosterol). Intriguingly, secretions from the armpit glands look like a brown paste containing almost exclusively squalene, a hydrocarbon terpenoid of low volatility. Male lemurs display a typical behavior: they rub their wrists to their armpits and secretions of these two glands are, therefore, mixed. Males then fix this mixture on a substrate, often a tree, using a typical wrist-marking behavior. In this way, males may signal their presence to possible competitors. The low volatility of the almost odorless paste produced by the armpits may bind more volatile compounds produced by the wrist glands [SCO 07a]. The final mixture and the associated chemical message could, therefore, last longer in the environment, even in the absence of the emitter. Surprisingly, this behavior is associated with a characteristic sound because males' wrists possess a nail that scarifies trees when scent-marking occurs. The resulting scar is visible in addition of being scented and audible when performed. Male lemurs, therefore, combine three communication modalities in one single behavior [SCO 07b].

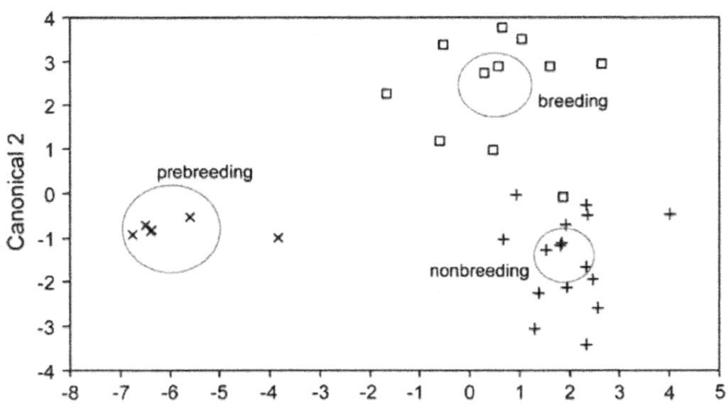

Figure 3.1. *Seasonal patterns in the composition of the secretions of the labial gland in female ring-tailed lemurs. The linear discriminant analysis allows distinguishing the pre-breeding (crosses), breeding (squares) and non-breeding (plus symbol) seasons (from [SCO 07a])*

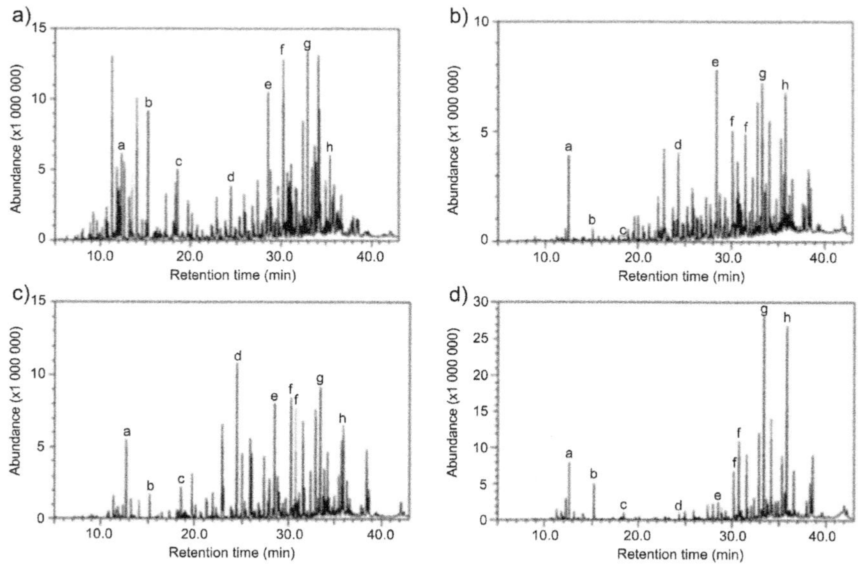

Figure 3.2. *Chromatograms of the scrotal secretions of two male ring-tailed lemurs. Top row: a male of high genetic diversity (neutral heterozygosity: 0.73) during a) the non-breeding season and b) the breeding season. Bottom row: a male of low genetic diversity (neutral heterozygosity: 0.21) during c) the non-breeding season and d) the breeding season. Letters identify the following compounds: a, internal standard (hexachlorobenzene); b, n-hexadecanoic acid; c, octadecanoic acid; d, octanoic acid, hexadecyl ester; e, squalene; f, isomers of tetradecanoic acid, tetradecyl ester; g, tetradecanoic acid, hexadecyl ester and h, hexadecanoic acid, hexadecyl ester (from [CHA 08])*

In ring-tailed lemurs, the proportion of the different odors produced by each of their four glands varies as a function of the season (Figure 3.1), suggesting that these odors may signal the reproductive state of the emitter. Mating patterns, and more broadly intra- and intersexual competition, depend largely on complex communication processes between individuals based on their body odors. In this species, individual scent signatures are also present in the secretions of the sexual glands indicating the identity of the emitter [SCO 07b]. In addition, recent studies have shown that scent signatures signal both males' and females' genetic quality: the most genetically diverse individuals present the most complex scents. Furthermore, in ring-tailed lemurs, genetically related individuals share more similar chemical profiles than unrelated individuals. Surprisingly, these chemical signals, encoding both individual's genetic quality and relatedness between individuals, are present during the breeding season only (e.g. [CHA 08]; Figure 3.2). At this time of the year, competition for reproduction is the highest; therefore, it should be beneficial to discriminate both individuals of high genetic quality (to reproduce with or to avoid competing with) and relatives (to avoid reproducing or competing with).

These chemical studies are corroborated by behavioral tests in controlled conditions (Figure 3.3). Again, these analyses illustrate how chemical communication is a major determinant of socio-sexual relationships in complex societies, such as those of lemurs, and its leading role in the evolution of sociality.

In contrast with lemurs, Old World primates, including great apes and humans, have long been considered as "microsmatic" species with an olfactory system that has been gradually reduced in favor of a more and more elaborate visual system. The underestimated role of olfaction in primate communication processes results from several factors. From an anatomical standpoint, the most widespread view is that, among mammals, Old World primates do not possess the vomeronasal organ (VNO) or possess only a vestige of such an organ. The VNO is specialized in sensing fluid-phase molecules, particularly non-volatile odors such as pheromones, thought to mediate reproductive activity. A typical mammalian-like VNO is clearly absent in most adult primates studied so far, including baboons, macaques and colobus monkeys [SMI 01]. There is, however, conflicting evidence: some adult humans and chimpanzees have retained an anatomical structure similar to a VNO (e.g. [SMI 01]) with, however, unknown functions. Moreover, the main olfactory epithelium, found in all mammalian species, including Old World primates, and rather specialized in the detection of small volatile molecules derived from numerous environmental sources, could share similar functions with the VNO, including the processing of pheromones.

Figure 3.3. *Behavioral test in ring-tailed lemurs. Scent secretions were rubbed on the two external wooden sticks (white arrows). The central wooden stick is a negative, unscented control (photo: Jeremy Chase Crawford, Christine Drea)*

From a genetic standpoint, most genes coding for pheromone receptors or genes playing a major role in VNO functions in rodents are non-functional pseudo-genes in Old World primates and humans (e.g. [DUL 95]). Some of these genes are, however, also pseudo-genes in marmosets, a New World primate characterized by a functional VNO. Thus, the different species studied so far could possess different sets of functional vomeronasal genes, which are not yet explored. Finally, recent empirical evidence suggests that Old World primates are morphologically and physiologically equipped for decoding socio-sexual cues contained in odors, a role that was previously attributed to the VNO and to species possessing it. The purported absence of an elaborate chemical communication system in these primates is called into question by: 1) the existence of specialized scent glands; 2) the presence of scent-marking behavior; 3) the production of known chemical signals; and 4) the perception of these signals by conspecifics (for review, see [CHA 13]). More specifically, in mandrills (*Mandrillus sphinx*) living in the humid equatorial forests of Central Africa, several pieces of evidence suggest a central role for olfaction in interindividual communication processes. First, this species is one of the rare Old World primates equipped with a sternal scent gland, present in both sexes. Second, secretions of this gland could provide socially pertinent cues (e.g. [SET 10]). This species also displays striking anatomical and behavioral specializations such as open nasopalatine ducts and flehmen behavior [CHA 13]. Surprisingly, these two specializations were only observed in mammal species with a functional VNO. In fact, flehmen, a typical behavior characterizing carnivores and ungulates (see Figure 3.3, Chapter 1), allows the transport of fluid-phase molecules to the VNO through nasopalatine or vomeronasal ducts. Further anatomical and physiological studies are, however, necessary to confirm that in mandrills, flehmen convey odors to a VNO-like structure through nasopalatine ducts.

The absence of systematic studies in anatomy, morphology, physiology and behavior has led to inaccurate generalizations on the role of chemical communication in primates, especially in Old World primates, including humans.

3.2. The role of odors in human communication

What about the sense of smell and taste in humans in the context of communication between individuals? What are our natural capacities to emit chemical messages and to receive and interpret them in a reliable way? These questions are not recent but they may provide new perspectives on the behavior and on the physiological functions of smell and taste in humans when viewed in the light of studies on mammalian communication.

3.2.1. *Human odors convey a large panel of cues*

Besides the simple exchange of information, chemical communication acts together with other senses to produce attentional, motivational or physiological changes in receivers who themselves initiate seeking or avoidance responses. Social interactions are not randomly distributed in social groups; therefore, the prime function of chemical communication is to allow interindividual recognition, underlain by notions of age, sex, parental status and group affiliation. Recognized individuals may be detected as motivated or not to interact, sexually receptive, stressed, in poor health or even as contagious. All these levels of individual recognition and individual biological condition are potentially transmitted by the main or accessory olfactory pathways in numerous mammals. As such, biologists seek for such recognition system in primates, especially in our own species.

The recognition of human conspecifics based only on their body odors was demonstrated when male and female judges were asked to determine the identity and sex of people. For example, when judges were asked to rate T-shirts worn by 10 different male and female donors, a significant proportion of judges (~30%) correctly identified the odor of their own partner [SCH 81]. Therefore, body odors carry relatively reliable information about individuality. Further, odors from palm sweat also allowed discriminating between sexes; the olfactory contrasts between man and woman were better discriminated than the contrasts between two women or two men, suggesting that assigning sex on the sole basis of odors was easier than assigning individuality. This sex differentiation of the axillary odors appeared to be based on a graded signal: the higher the intensity of the scented stimulus, the more it was evaluated as unpleasant and masculine.

Individuality in body odors depends on environmental and genetic processes that engender complex olfactory signatures in terms of both chemistry and semantics. For example, adult judges were capable of detecting olfactory cues of parenthood by matching mothers' odors to those of their infants or odors of twins. These judges may, therefore, detect a general olfactory proximity between individuals, a family odor, and also a particular individual's olfactory characteristics linked to lifestyle such as diet or age. For example, when monozygotic twin sisters shared the same diet, they were more difficult to discriminate from each other based on their palm odors than unrelated women were. Moreover, these odors were less well differentiated in twins sharing the same diet than in twins with contrasted diets.

Genes of the major histocompatibility complex (called *Human Leucocyte Antigens*, HLA, in humans) illustrate the genetic basis of similarities between olfactory phenotypes. These HLA genes are involved in the immunological response to pathogens, they are highly polymorphic and, therefore, constitute potential markers of olfactory individuality. The immunogenetic type influences body odors, as indicated by experiments during which women were asked to evaluate T-shirts worn by men of known HLA genotype. Three males' genotypes were genetically dissimilar to females' genotypes and three were similar. When not using the contraceptive pill, women rated male odors as more pleasant when their genotypes were dissimilar compared to their own. A similar result was found when men had to express their preference for the odor of T-shirts worn by women [WED 97]. Thus, genetic factors may simultaneously influence chemical profiles of cutaneous secretions (carried in native secretions or due to the action of the skin microbiota) and olfactory mechanisms that direct preferences toward dissimilar genotypes.

To summarize, body odors may convey different categories of cues that are "easily" decoded in laboratory conditions. These chemical messages result from gene–environment interactions specific to each individual, but their perception is likely to fluctuate as a function of the internal and external contexts of the individual receiver.

3.2.2. *Body odors reflect internal states*

Body odors reflect fluctuations in physiological states. Their chemical exteriorization has been studied in women, for example, during their fertile phase. Male judges were able to discriminate T-shirts worn by women during their follicular or their luteal phase (before and after ovulation, respectively), and evaluated the first as producing more pleasant and more sexually stimulating odors [SIN 01]. Emotional states are another cause of intra-individual variability in body odors. Axillary sweat associated with experiences of fear or anxiety was different from sweat associated with neutral experiences. The odor of anxiety was perceived as more intense, less pleasant and more suggestive of aggressiveness. In addition, sweat produced during anxious periods affected the emotional state of the receiver by raising his/her perceived anxiety, by reducing the threshold of defensive reflexes, by reducing pre-attentional treatment of danger from the environment, by accelerating choice decisions in emotionally ambivalent situations and, finally, by facilitating risk taking in ambiguous situations (reviewed in Schaal [SCH 13]).

Anxiety and fear convey additional chemosensory correlates. For example, the inhalation of the odorless scent of female tears of sorrow by men decreased their sexual attraction for female faces, reduced men's psychophysiological arousal, lowered testosterone levels and diminished neural responses in brain areas that are typically activated by erotic images [GEL 11]. Female tears of sorrow could thus carry inhibitor agents to the emotions underlying sexual arousal and aggression. By contrast, axillary sweat sampled during euphoric times (e.g. following a victorious sports match) increased the amplitude of the electrodermal response, an index of attentional arousal [ADO 10]. Finally, the analysis of activity patterns of the brain area dealing with emotions showed that the emotional experience of sexual nature generated axillary secretions with different odors from those linked to neutral stimulations [ZHO 08].

Future studies should elucidate processes underlying the production and emission of the volatile message(s) (histological sources and chemical identity; kinetics of odor discharge; relationships between nature/intensity of stress and emitted odors; effect of donor's sex) as well as processes of odor perception (automaticity and universality of responses; effect of receiver's sex).

3.2.3. *What are the functions of social smells in human daily life?*

How chemical cues, decoded in the laboratory, are perceived *in vivo* and what are their utilities? Olfactory cues of individuality, sex, physiological or emotional states operate in intimate contexts. Males and females use, however, their olfactory capacities in different ways. Women pay more attention to the scent of their sexual or romantic partner when asked to weigh olfactory, tactile, auditory or visual stimuli. By contrast, men assign as much weight to visual traits as to olfactory stimuli from their partner. During heterosexual activity, men declare to rely on their visual, auditory and tactile senses while women rely more on their olfactory sense [HER 97]. Attraction for olfactory genotypic traits and the function of odors in sexual selection and sexual physiology both appear meaningful in this context.

The influence of odors in sexual choices was also addressed indirectly when analyzing matrimonial relationships in Anabaptists of Pennsylvania, a North-American endogamous isolate [OBE 97]. In this community, both partners of 411 couples were genotyped at their HLA genes. This population carries a relatively limited number of different HLA alleles because of endogamous mating since its foundation in the 16th Century. The distribution of marriages in this population did not match a simulated distribution based on random choices of mating partners, indicating that chance alone did not explain marriages. By contrast, HLA genotypes appeared to be major determinants of partner choices. Individuals with more dissimilar alleles had a higher probability of getting married (or remaining married) than more genetically similar partners. This result corroborates those reported above where body odors produced by immunogenetically dissimilar persons were preferred.

Parent–infant relationships constitute another system where human olfaction plays a significant role. Parents are particularly discerning with their infant's body odors: just after birth, newborns are examined under all traits, including their odors. Their olfactory individuality is thus acquired within the first hours or days following birth. Smelling newborns may reflect a particular state of olfactory reactivity from post-parturient mothers and an infant's odor may influence the initial development of mother–infant attachment. The emotional disinvestment of infants who suffer from pathologies affecting their odors or, in less extreme cases, when infant's odors are temporarily modified, demonstrate the importance of the olfactory bases of parental ties in humans. Reciprocally, maternal odors strongly influence infant's behavior. Mammary secretions regulate the behavior of newborns who tend to calm down when exposed to breast odors and orient their searching actions as soon as the first minutes following birth. This directional effect of maternal odors was used to identify some mammary secretions which were particularly reactogenic for newborns ([DOU 09]; see Figure 3.4) and to analyze the degree of infant's olfactory selectivity as a function of prenatal and postnatal olfactory experience. These initial olfactory preferences initiate infant–mother attachment and, by their aptitude to mobilize visual attention, promote an infant's acquisition of the multisensory identity of parents [DUR 13].

Body odors also regulate social proximity outside the family. From 4 to 5 years of age, children recognize body odors of schoolmates, and girls appear better both at discriminating and at being discriminated than boys. In children, these group dynamics often indicate the affiliative networks and the importance of being olfactorily compatible with others. Stigmatization of deviant body odors easily emerge in these contexts. It is by identifying this common sensory part that most human groups form internal structure and stand out, or allege to stand out, from their neighbors.

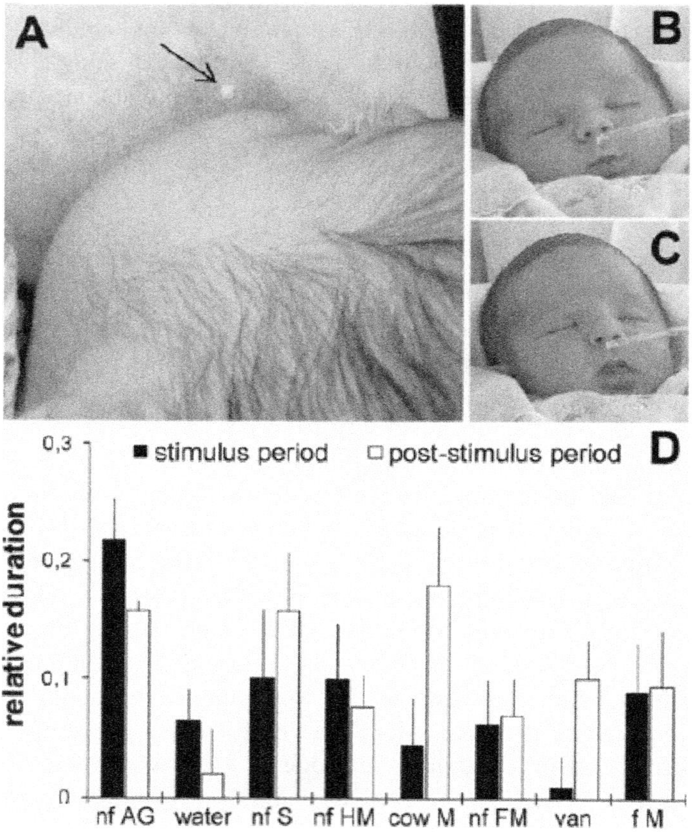

Figure 3.4. *Human areolar glands and neonate behavior: the odors of areolar secretions elicit longer duration of oral/facial responsiveness than any other reference stimuli. A) Areola of a lactating woman with Montgomery's glands producing visible secretion (arrow); B and C) oral response to the secretion of Montgomery's gland (B: lip pursing, C: tongue protrusion) in a sleeping 3-day-old infant. D) mean relative durations of neonates' (n = 19) oral responses during a 10 stimulation period (black bars) and a 10 s period after stimulation (empty bars) with several olfactory stimuli (AG: areolar glands; S: sebum; HM: human milk; cow M: cow's milk; FM: formula milk; van: vanillin; M: milk; f: familiar; nf: non-familiar (from [DOU 09])*

The chemistry of olfactory cues, probably conveyed in complex secretions and involved in the performance of recognition summarized above, remains unknown. The odors of the human body are composed of at least 350 volatile compounds, the combinations of which indicate individuality, sex, emotional state or health [DOR 13]. Environmental factors (related to feeding diet or skin microflora) or shared genetic characteristics may generate shared odor signatures across individuals. Along with these complex chemical signatures, we cannot yet exclude that crucial, universal, compounds shared by different species may also convey critical information.

3.2.4. *Human pheromones, fact or fiction?*

In rare cases, supra-individual odors (e.g. chemical molecules found in all newborns or in all mature males from the same species) have been identified and their unconditional signal evaluation demonstrated within a same species. These signals are called "pheromones". At first considered in an insect, the mulberry moth, the notion of the pheromone has been applied to any substance, volatile or not, emitted by an individual and perceived by a conspecific in which it elicits a definable behavioral or physiological response (see Chapter 1; for review, see [WYA 14]). Such pheromonal signals trigger physiological or behavioral automatisms, which are, in theory, innate. In other words, their effects do not depend on prior exposure or learning effects.

The assumption that some molecules constituting human body odors might operate like pheromones results from the hasty generalization of mechanisms described in other mammals to our own species. Mammalian pheromones were initially considered in sexual selection studies; they were, therefore, first explored in the *Homo* genus within this theoretical context. For example, a mixture of five fatty acids emitted in vaginal secretions of female rhesus macaques (called "copulins") were also detected in secretions of human females, but their behavioral effect has never been demonstrated. Additionally, the two steroids, androstenone and androstenol, which give to human axillary sweat its urinous or musky qualities, have attracted many studies because both molecules are emitted by the boar and induce postures indicating sexual receptivity in the sow. By contrast, the behavioral effects of these two steroids in humans were either absent, weak or highly variable across the different behavioral tests performed. Finally, other steroids isolated from natural secretions of the human skin (androstadienone) by the perfume industry or synthesized *de novo* (pregnadienedione, estratetraenol) were called "vomeropherines" because they were supposed to have a strong affinity for hypothetical receptors of the VNO. This organ seems, however, to have no function in adult humans. Androstadienone appeared rather to modulate autonomic nervous system responses (electrodermal reaction) and mood states in both women and men, but studies were often either contradictory or non-reproducible.

A chemosensory mediation of the synchronization of ovarian cycles in women who lived together for an extended period of time was also explored [STE 98]. However, these experiments have raised many concerns including methodological or statistical issues [SCH 01]. The so-called pheromonal influences have also been explored in male sexual physiology. For example, a stimulating effect of the androstadienone molecule on the circulating level of testosterone was suggested. The uniqueness of these compounds in eliciting physiological effects is still, however, awaiting rigorous confirmation. Brain imaging appears to be a good way to decipher the cerebral regions involved in these hormonal changes but, besides the fact that such steroidal odors can be active at a subthreshold level, almost everything remains yet to be done in this area of research.

This brief survey on chemical communication in humans highlights the difficulty in analyzing behavioral and physiological consequences of natural olfactory stimulations in our own species and, therefore, in establishing the reliability and the value of the concept of pheromones in humans. It seems obvious that human responses to conspecific odors are far from having the same functional clarity compared with sexual or parental pheromones described in other mammals. Human responses to homologous substances have rarely been dissociated from the effects of prior experience and from the context in which they are perceived. Currently, the concept of pheromones which involves unconditional responses, varying only slightly in a given context and supported by pre-wired perceptual mechanisms, has not found any convincing chemical candidate in humans [WYA 15].

These observations regarding pheromones are not conflicting with the view that humans have elaborate abilities to perceive and use conspecific body odors in several contexts of interindividual communication. By contrast, the ancient use of perfumes is now studied as a domain of interactions between biology and culture shaping human chemical communication, and not only exclusive of natural body odors.

3.3. The senses of smell and taste in the search for food and remedies

Sensory ecology, a particular development of chemical ecology, concerns the role of organoleptic perceptions in the functioning of a species. Odors are, in numerous species including primates, the support of fundamental interindividual communication processes, although discrete and often unconscious. When this is associated with taste (forming "flavors"), they also support multiple interactions between individuals and their vegetal or mineral environments. Smell and taste have also evolved in animals depending on their needs to feed, protect and care for. Here, we address these fields, going from the relationship between senses and nutrition in primates to selection/avoidance relationships leading to self-medication in animals and human medicines. In all these domains, the role of markers, if not mediators, of allelochemical molecules will be the common thread.

3.3.1. *Interactions between senses and food in primates*

As proposed by McKey and colleagues [MCK 81], food selection can be approached as a cost-benefit problem in order to optimize energy intake. To seek sources of energy in molecules, such as sugars, or to avoid indigestible (e.g. polyphenols, tannins) or toxic (e.g. alkaloids, saponosides) metabolites, the sensory systems of animals are adapted to the detection of more or less complex small molecules. With the exception of sugars, which are part of the primary metabolism, most odorous or tasty molecules correspond to the so-called secondary metabolites. These latter molecules have been selected for their physiological properties (e.g. repulsive or toxic for predators, or by contrast, attractants for pollinators or seed dispersers; see Introduction and Chapter 1). Secondary metabolites offer a selective advantage to some plant species in terms of survival and their detection is an important issue in animals. For example, the threshold of glucose perception is lower in primate species with a frugivorous diet. In humans, the threshold of glucose and saccharose perception is likewise more elevated in African forest societies, where fruits are abundant (Twa, Aka and Gieli Pygmies), than in savannah societies (Koma, Doupa), where fruits are rarer and where their detection is more advantageous. Both in primates and in humans, this mechanism is associated with a hedonic component, as the intake of sugar is associated with pleasure. If these eating modalities result from an adaptive process, it is the same for sensory perceptions.

Other sensations characterize the gustatory environment of mammals: astringency, pungency, bitterness, saltiness or acidity. Bitterness is due to several classes of allelochemical substances with alkaloids playing a major role. These molecules are often bioactive and include some of the most toxic compounds of the plant kingdom (albeit, all bitter molecules are not equally toxic). The ability to detect such molecules is, therefore, highly associated with the tolerance/resistance of the animal. Ruminant herbivores (grazers) that are rarely exposed to these molecules are thus more sensitive than herbivores that rely more on diversified resources (browsers). Tannins are complex molecules considered for a long time to be defense substances. Sensitivity to these molecules appears to be higher in grazer herbivores for which diet is rarely rich in tannins than for browsers that are constrained to eat them regularly. The production of proteins aiding detoxification is also higher in the latter species.

In primates, mechanisms of avoidance of toxic substances seem to be linked to complex learning mechanisms [MAS 12], as well as to inherited phenomena, as evidenced by the gusto-facial reflex. This reflex corresponds to (reproducible) facial responses produced in reaction to gustatory stimuli and is already observed in newborns of both human and non-human primates.

3.3.2. *Senses and self-medication in animals*

In addition to nutritional behaviors, flavors are involved in medical practices. The gap between food and therapy is, however, relatively artificial. Ingestion of vegetal substances satisfies food requirements as well as preventing some pathologies or re-establishing altered physiological functions. Animal self-medication, or zoopharmacognosy, has been demonstrated in numerous species, from great apes to ruminant mammals, as well as in birds or insects. For some of these species, individuals discern some categories, if not some plant species, for therapeutic purposes [MAS 12]. The therapeutic consumption of plants is of two types: the prophylactic type, with a more or less regular consumption of plants containing anti-parasitic secondary metabolites, or the reactive type, with a transitory consumption responding to a given pathological condition [HAR 05]. The latter form is the most intriguing one because it involves an implicit selection mechanism, while prophylaxis may be considered as a long-term selection mechanism. For example, neophobia (aversion to unusual foods), a frequent trait found in mammals, tends to neophilia during parasitic episodes, provoking diet diversification and thus probably partially explaining the basis of self-medication. In great apes, Masi *et al.* [MAS 12] also highlighted different feeding behaviors between the homeostatic state and pathological episodes. These differences are more pronounced in chimpanzees than in western lowland gorillas possibly because chimpanzees are more frugivorous than gorillas, and their capacity to detoxify secondary metabolites is thus lower. The impact of these unusual food consumptions appears more important in chimpanzees, leading to a higher selective advantage in this species. Along with these observations, bitterness also seems to be an important taste, because it is presumably sought by chimpanzees during pathological states [KRI 05]. The study of chemical diversity patterns among different plants consumed would probably help understanding food selection criteria more precisely.

The case of capuchin monkeys is of particular interest, because of their topical use of arthropod species. These monkeys rub millipedes on their fur for the supposed purpose of repelling mosquitoes that are vectors of parasitic worms [VAL 00]. These arthropods produce two highly odorous molecules (benzoquinone and cresol) and this prophylactic behavior is probably linked to their scent. This brings us back to the question of senses, because these behaviors appear to be likely based on the organoleptic properties of plants (and, more rarely, of insects or invertebrates) and their perceptions by animals. One next step in chemical ecology would be to understand which cognitive mechanisms are elicited by these odorant or tastant molecules as a function of the animal species.

3.3.3. *Senses in human therapies*

These behaviors observed in animals are also found in human societies: many plants used by chimpanzees for medicinal purposes are known for their therapeutic properties in African human societies [KRI 05]. In Amerindian societies, tastes and odors are also used to decipher the medicinal properties of plant species [LEO 02].

Figure 3.5. *Left: Teko Amerindian of French Guiana identifying a tree thanks to its odor, on an experimental plot in the forest. Right: wood goblet of Quassia amara L. is acclaimed by Surinamese Saramaka, for the extreme bitterness that the wood transfers to its contents, water or rum. These goblets are used as tonics, digestives or even anti-malarial medicine, and are linked to the Afro-surinamese cultures (photo: G. Odonne)*

It is now widely recognized that ethnopharmacopoeias are not only collections of randomly selected plants. Some botanical families are, indeed, more represented than others, independently of their abundance in the local floras. Large-scale comparisons show that bioactive metabolite-producing families tend to be overrepresented. Indeed, bioactive secondary metabolites are often molecules showing distinctive organoleptic properties. Most of the small terpenes are volatile and aromatic, numerous alkaloids as well as cyanogenic glycosides are bitter and tannins are astringents. For the Yucatec Maya, bitter plants are mostly used for skin problems and for bites/stings from venomous animals, while they are generally preferred to relieve pain in the Popoluca [LEO 02]. In North-East Brazil, bitter plants are favored for indigestion, inflammation and influenza. Van Andel and colleagues [VAN 12] showed in societies of African origin, living on both banks of the Atlantic Ocean, that bitterness is associated with aphrodisiac properties, malarial disease, respiratory infections and genito-urinary problems. The relationship between sensory perceptions and therapeutic activities is, therefore, logical and no longer debated [LEO 02].

3.3.4. *An evolutionary conception of the link between senses and health*

In humans, bitter taste is detected by G protein-coupled receptors coded by a gene family called TAS2Rs. At a global geographical scale, important differences in the polymorphism of one of these genes (TAS2R16) are found across populations. Moreover, in some regions of Africa, alleles coding for less sensitive receptors are more frequent. A positive correlation between a higher prevalence of malaria and a lower sensitivity to bitterness might be established. Natural agonists of this receptor are relatively ubiquitous bitter molecules, the β-glucopyranosides. Interestingly, at sublethal and chronic doses, these molecules show a relative anti-plasmodial effect. This complex relationship between humans, plants and *Plasmodium*, responsible for malaria, is also particularly interesting because it involves a chemical mediator detected by human sensory organs [ETK 03]. We could also hypothesize that African societies make much of bitter sensations [VAN 12] because of a strong environmental determinism on this behavior.

Another example of tripartite interactions concerns spices and condiments. In traditional culinary preparations, spice load increases proportionally with the average temperature of the concerned regions (an elevated temperature reduces the duration of conservation of food; [BIL 98]). Indeed, most spices show anti-bacterial properties. Thus, by reducing contamination, spices help in avoiding digestive diseases, hence providing a selective advantage to individuals who find their taste pleasant [BIL 98]. If cooking is an eminently cultural fact, there is also an environmental determinism on these behaviors.

Plant-based therapies originate from an evolutionary model, refined and expanded by learning and culture [HAR 05]. The use of biodiversity surrounding human beings for medical purposes is an approach originating from food avoidance/selection relationships, as well as informed self-medication behaviors observed in some animal species. These behaviors are nevertheless rarely integratively studied; a solid interpretation, therefore, remains difficult. Nonetheless, it seems clear that environmental mechanisms have a high impact on medicinal/feeding behaviors, corresponding to cultural features in humans. Sensory ecology is at a crossroad with several physiological mechanisms, responding to a given environmental determinism and cultural phenomena. Chemical ecology, by studying the origins of these mechanisms via mediatory molecules, will help to elucidate many shadow areas.

Figure 3.6. *Steam bath in the Yanesha, Peru. This treatment allows the extraction of volatile and aromatic terpenes from plants (photo: Valadeau)*

3.4. Conclusions – the adaptive functions of the sense of smell in "microsmatic" species

Throughout the evolutionary history of primates, olfaction has contributed advantageously to the organization of individuals' daily activities and of social groups. Odors produced by natural objects, in particular food and medicinal substances, generated sophisticated knowledge (e.g. pleasant/unpleasant, edible/toxic, usable or not as cosmetics) and differentiated uses according to the local ecological resources. Non-human and human primates thus rely on their sense of smell (also on taste and on odor-taste combinations) to locate, recognize and select their food resources and to transform and conserve them. As demonstrated in the third part of this chapter, olfaction is also widely involved in hygienic practices (directed to the body and the domestic spheres), and in therapeutic and defensive uses (against predators and parasites). Odors used as repellent for parasites (found in several primates) and floral or fruity odors used as attractants for conspecifics have probably played an evolutionary role in the emergence of hedonic and aesthetic evaluation of odors which, worn as ornaments in humans, have warranted individual's well-being in the context of social conformity. Such odors could also manipulate conspecifics or induce the synchronization of collective emotions. In the long run of biocultural evolution, all these practices attest to the fundamental awareness and universal attention to the odor world in primates in general, and in our own species in particular.

It may, however, appear paradoxical to investigate the sense of smell in species classified, in the 19th Century, as "anosmatic" animals, for which the sense of smell was considered a sensory relic of evolutionary processes, which would have over-invested in vision and audition. This "anosmatic" class of animals comprised primates (in particular great apes and humans), as well as monotremes, Tubulidentata, cetaceans and birds, in opposition to fishes, reptiles, rodents, carnivores and other ungulates classified as "osmatic" animals. A later classification acknowledged some remnants of olfactory capabilities in primates, which were then re-qualified as "microsmatic". While based on undeniable morphological and anatomical criteria, this classification somehow reflects scientific stereotypes imprinted from anthropomorphism, and assigns to animals a sensory dominance inferred from structural appearances rather than from effective perceptual performances [SCH 81]. This has been recently reinforced with data from molecular biology that showed a correlation between the number of genes coding for olfactory receptors and capacities of detection/discrimination of odors (e.g. [ROU 00]). If the human olfactory tract is indeed quantitatively minimized, in surface, volume and number of olfactory receptors, we cannot omit that this evolution occurred with the maximization of a neocortical "supercomputer" capable of extreme analytical detail as well as multisensory integration. This cognitive amplification of olfaction in primates has been accompanied by important cultural amplification characterizing our own species: the aptitude to translate olfactory impressions into mental images, concepts and words, which facilitate the transmission of related emotions and knowledge. More generally, studies reviewed in this chapter are clearly in opposition to the notion of primates being functionally "microsmatic" and underline the utility of revisiting old paradigms to envision new fields of research.

3.5. Bibliography

[ADO 10] ADOLPH D., SCHLÖSSER S., HAWIGHORST M. *et al.*, "Chemosensory signals of competition increase the skin conductance response in humans", *Physiology & Behavior*, vol. 101, pp. 666–671, 2010.

[BIL 98] BILLING J., SHERMAN P.W., "Antimicrobial functions of spices: why some like it hot", *Quarterly Review of Biology*, vol. 73, pp. 3–49, 1998.

[CHA 08] CHARPENTIER M.J.E., BOULET M., DREA C.M., "Smelling right: the scent of male lemurs advertises genetic quality and relatedness", *Molecular Ecology*, vol. 17, pp. 3225–3233, 2008.

[CHA 13] CHARPENTIER M.J.E., MBOUMBA S., DITSOGA C. *et al.*, "Nasopalatine ducts and flehmen behavior in the mandrill: reevaluating olfactory communication in Old World Primates", *American Journal of Primatology*, vol. 75, pp. 703–714, 2013.

[DEL 12] DELBARCO-TRILLO J., SACHA C.R., DUBAY G.R. *et al.*, "Eulemur, me lemur: the evolution of scent-signal complexity in a primate clade", *Philosophical Transactions of the Royal Society of London B*, vol. 367, pp. 1909–1922, 2012.

[DOR 13] DORMONT L., BESSIÈRE J.-M., COHUET A., "Human skin volatiles: a review", *Journal of Chemical Ecology*, vol. 39, pp. 569–578, 2013.

[DOU 09] DOUCET S., SOUSSIGNAN R., SAGOT P. *et al.*, "The secretion of areolar (Montgomery's) glands from lactating women elicits selective, unconditional responses in neonates", *PLoS ONE*, vol. 4, p. e7579, 2009.

[DUL 95] DULAC C., AXEL R., "A novel family of genes encoding putative pheromone receptors in mammals", *Cell*, vol. 83, pp. 195–206, 1995.

[DUR 13] DURAND K., BAUDOUIN J.Y., LEWKOWICZ D. *et al.*, "Eye-catching odors: familiar odors promote attention and sustained gazing to faces and eyes in 4 month-old infants", *PLoS ONE*, vol. 8, p. e7067, 2013.

[ETK 03] ETKIN N.L., "The co-evolution of people, plants, and parasites: biological and cultural adaptations to malaria", *Proceedings of the Nutrition Society*, vol. 62, pp. 311–317, 2003.

[FRE 12] FREEBERG T.M., DUNBAR R.I., ORD T.J., "Social complexity as a proximate and ultimate factor in communicative complexity", *Philosophical Transactions of the Royal Society of London B*, vol. 367, pp. 1785–1801, 2012.

[GEL 11] GELSTEIN S., YESHURUN Y., ROZENKRANTZ L. *et al.*, "Human tears contain a chemosignal", *Science*, vol. 331, pp. 226–230, 2011.

[HAR 05] HART B.L., "The evolution of herbal medicine: behavioural perspectives", *Animal Behaviour*, vol. 70, pp. 975–989, 2005.

[HER 97] HERZ R., CAHILL E., "Differential use of sensory information in sexual behaviour as a function of gender", *Human Nature*, vol. 8, pp. 275–286, 1997.

[KRI 05] KRIEF S., HLADIK C.M., HAXAIRE C., "Ethnomedicinal and bioactive properties of plants ingested by wild chimpanzees in Uganda", *Journal of Ethnopharmacology*, vol. 101, pp. 1–15, 2005.

[LEO 02] LEONTI M., STICHER O., HEINRICH M., "Medicinal plants of the Popoluca, México: organoleptic properties as indigenous selection criteria", *Journal of Ethnopharmacology*, vol. 81, pp. 307–315, 2002.

[MAS 12] MASI S., GUSTAFSSON E., SAINT JALME M. *et al.*, "Unusual feeding behavior in wild great apes, a window to understand origins of self-medication in humans: role of sociality and physiology on learning process", *Physiology & Behavior*, vol. 105, pp. 337–349, 2012.

[MCK 81] MCKEY D.B., GARTLAN J.S., WATERMAN P.G. *et al.*, "Food selection by black colobus monkeys (*Colobus satanas*) in relation to plant chemistry", *Biological Journal of the Linnean Society*, vol. 16, pp. 115–146, 1981.

[OBE 97] OBER C., WEITKAMP L.R., COX N. *et al.*, "HLA and mate choice in humans", *American Journal of Human Genetics*, vol. 61, pp. 497–504, 1997.

[ROU 00] ROUQUIER S., BLANCHER A., GIORGI D., "The olfactory receptor gene repertoire in primates and mouse: evidence for reduction of the functional fraction in primates", *Proceedings of the National Academy of Sciences*, vol. 97, pp. 2870–2874, 2000.

[SCH 81] SCHLEIDT M., HOLD B., ATTILI G., "A cross-cultural study on the attitudes towards personal odors", *Journal of Chemical Ecology*, vol. 7, pp. 19–31, 1981.

[SCH 91] SCHAAL B., PORTER R., "“Microsmatic humans” revisited: the generation and perception of chemical signals", *Advances in the Study of Behavior*, vol. 20, pp. 135–199, 1991.

[SCH 01] SCHANK J.C., "Measurement and cycle variability: reexamining the case for ovarian-cycle synchrony in primates", *Behavioural Processes*, vol. 56, pp. 131–146, 2001.

[SCH 13] SCHAAL B., "Comment les émotions se transforment en odeurs? Vers l'hypothèse de la contagion émotionnelle olfactive", in SCHAAL B., FERDENZI C., WATHELET O. (eds), *Odeurs et Emotions. Le Nez a ses Raisons…*, Editions Universitaires de Dijon, Dijon, pp. 293–335, 2013.

[SCO 07a] SCORDATO E.S., DREA C.M., "Scents and sensibility: information content of olfactory signals in the ringtailed lemur (*Lemur catta*)", *Animal Behaviour*, vol. 73, pp. 301–314, 2007.

[SCO 07b] SCORDATO E.S., DUBAY G., DREA C.M., "Chemical composition of scent marks in the ringtailed lemur (*Lemur catta*): glandular differences, seasonal variation, and individual signatures", *Chemical Senses*, vol. 32, pp. 493–504, 2007.

[SET 10] SETCHELL J.M., VAGLIO S., MOGGI-CECCHI J. *et al.*, "Chemical composition of scent-gland secretions in an Old World monkey (*Mandrillus sphinx*): influence of sex, male status, and individual identity", *Chemical Senses*, vol. 35, pp. 205–220, 2010.

[SIN 01] SINGH D., BRONSTAD P., "Female body odor is a potential cue to ovulation", *Proceedings of the Royal Society of London B*, vol. 268, pp. 797–801, 2001.

[SMI 01] SMITH T.D., SIEGEL M.I., BHATNAGAR K.P., "Reappraisal of the vomeronasal system of catarrhine primates: ontogeny, morphology, functionality, and persisting questions", *The Anatomical Record*, vol. 265, pp. 176–192, 2001.

[STE 98] STERN K., MCCLINTOCK M., "Regulation of ovulation by human pheromones", *Nature*, vol. 392, pp. 177–179, 1998.

[VAL 00] VALDERRAMA X., ROBINSON J.G., ATTYGALLE A.B. *et al.*, "Seasonal anointment with millipedes in a wild primate: a chemical defense against insects?", *Journal of Chemical Ecology*, vol. 26, pp. 2781–2790, 2000.

[VAN 12] VAN ANDEL T., MITCHELL S., VOLPATO G. *et al.*, "In search of the perfect aphrodisiac: parallel use of bitter tonics in West Africa and the Caribbean", *Journal of Ethnopharmacology*, vol. 143, pp. 840–850, 2012.

[WED 97] WEDEKIND C., FÜRI S., "Body odour preferences in men and women: do they aim for specific MHC combinations or simply heterozygosity?", *Proceedings of Biological Sciences*, vol. 264, pp. 1471–1479, 1997.

[WYA 14] WYATT T.D., *Pheromones and Animal Behaviour. Chemical Signals and Signatures*, 2nd ed., Cambridge University Press, Cambridge, 2014.

[WYA 15] WYATT T.D., "The search for human pheromones: the lost decades and the necessity of returning to first principles", *Proceedings of Biological Sciences*, vol. 282, p. 2014–2994, 2015.

[ZHO 08] ZHOU W., CHEN D., "Encoding human sexual chemosensory cues in the orbitofrontal and fusiform cortices", *The Journal of Neuroscience*, vol. 28, pp. 14416–14421.

4

Microbiota and Chemical Ecology

4.1. The protagonist microorganisms of chemical ecology

Microorganisms – prokaryotes (bacteria, cyanobacteria and archaea) or eukaryotes (filamentous fungi, protists) – live in communities where there is a strong competition for occupying a niche and nutrient consumption. According to biotic or abiotic environmental constraints, the microorganisms produce an arsenal of molecules that they use as mediators of signaling processes or as effectors of microbial competitions in a given biotope, whether in a terrestrial or a marine environment. These compounds are molecules of quite variable chemical nature and molecular weight, small molecules, peptides or proteins which are produced from elementary bricks through sophisticated enzymatic complexes, organized in metabolic networks.

Thus, for example, "quorum sensing", the mechanism by which bacteria regulate their growth as a function of the density of the diverse populations present, is governed by the emission of signaling molecules, such as acyl-homoserine lactone or peptides. On the other hand, lactic acid bacteria are capable of producing powerful antimicrobial peptides (AMPs) (see also Chapter 9), which constitute a chemical arsenal to fight against invasion from other microorganisms in the microbiota.

Besides these intramicrobiota interactions, certain microorganisms live in association with multicellular eukaryote organisms and establish a chemical dialog by means of the reciprocal exchange of message molecules. These microorganisms associated with their host could, therefore, maintain diverse

---

Chapter written by Soizic PRADO, Catherine LEBLANC and Sylvie REBUFFAT.

relationships, from symbiosis to parasitism, which are mostly mediated by the bilateral production of compounds. It is in this way that plants have established intimate relations with fungi and bacteria, which are essential for reciprocal development of the plant and its microbial partners. Similar associations have been observed in algae, insects and mammals (see Introduction). The interconnection of these relationships leads to the recent emergence of the holobiont concept, which encompasses, as a subject of study, the host and its range of associated microorganisms.

In addition, this chemical communication, present at every interaction level of a microbiota, is genetically finely regulated by numerous stimuli in the ecosystems (stress, signal molecule, defense molecule, etc.) leading to modifications of metabolic production. The knowledge of their mechanisms of production (biosynthesis) and their modeling (molecular networks) is therefore required not only for understanding these microbial interactions, but also for their applications for environmental, ecotechnological and biotechnological or medical purposes (see Chapter 9).

4.2. Strategies for the study of microbiota

The molecular dialog between an organism and its associated microbiota, that is, in the holobiont (see section 4.4.4.2 and Chapter 1), is based on numerous actors and a large range of chemical interactions. To understand this complexity, many strategies, often complementary, are used.

4.2.1. *How should the microbiota be characterized?*

A first strategy, called culture-dependent, consists of isolating the constituents of the microbiota by developing adapted culture methods. This approach helps to study separately the host and microbiota, classifying the microorganisms in the laboratory at the genetic level (complete genome sequencing), physiologically and biochemically, and developing functional approaches to test the role of the microorganisms in the interaction (Figure 4.1). Meanwhile, it presents an important limit as only the microorganisms that can be cultured and isolated are taken into account by this approach.

The advent of high-throughput DNA sequencing techniques (Next-Generation Sequencing Technology or NGS, see Chapter 7) has since helped to understand the real biodiversity present inside the microbiota, as illustrated in the studies of intestinal microbiota of mammals (culture-independent methods). These approaches, complementary to the culture methods, rely on the DNA sequencing extracted from the surface microbiome or from the holobiont in its entirety, by targeting a marker gene like 16S or 18S rRNA (metabarcoding), or integrally (metagenomics). The computer processing of the sequences thus obtained helps to identify taxonomic groups by sequence homologies (Operational Taxonomic Unit or OTU) to describe the bacterial or fungal phylogenic diversity of the microbiota (metabarcoding data), or to reconstruct genomes of microorganisms associated with a host (metagenomic data (see Chapter 6)).

4.2.2. *What tools are available to help understand the roles of the microbiota?*

One of the challenges in chemical ecology is to decipher the molecular bases of the interaction between the microbiota and its host. It is possible to apply "omics" approaches to the different partners in interaction or to characterize the whole range of transcribed genes (transcriptomics) and expressed proteins (proteomics) separately (see Chapter 6). Comparative analyses are then performed to search for genes or proteins potentially involved in the communication processes or their regulation. Complementary to these transcriptomic and proteomic approaches, metabolomics represent analytical resources of choice to identify the molecule(s), responsible for a given interaction between two organisms, because these signals are often present in very small quantities among the other naturally produced molecules (see Chapter 7). Based on the use of nuclear magnetic resonance (NMR) or mass spectrometry (MS) combined with the separation by gas (GC–MS) or liquid (LC–MS) chromatography, these techniques help to analyze the range of synthesized metabolites and the evolution of the chemical content of each of the players or of their exudates. The comparison of metabolomes or exo-metabolomes obtained under different conditions helps identify potential chemical signals involved in the studied interactions. The identification of these chemical signals, often a delicate step, could thus be ensured by querying "in house" or published databases (e.g. Dictionary of Natural Products, SciFinder, MarinLit). In the case of absence of correlation, it involves the preliminary isolation of the compounds of interest in view of their complete classification by spectroscopic methods (NMR and MS) (see Chapter 8).

Combining metabolomics and metagenomics has helped explore the functioning of certain bacterial communities (soil, digestive tract) and progress in the comprehension of bacterial aggregates (biofilms). These global approaches also apply to the emergence of powerful bioinformatic tools, capable of integrating all generated data, such as the reconstruction of metabolic networks *in silico* from genomic data issued from the host, microbiota or holobiont. The following step consists of analyzing the metabolic flux following enrichment of cultures with stable isotopes like carbon-13 (^{13}C). These experimental fluxomic data help to implement the metabolic network models for a better understanding of the microbial metabolic dynamics in interaction with the host (Figure 4.1).

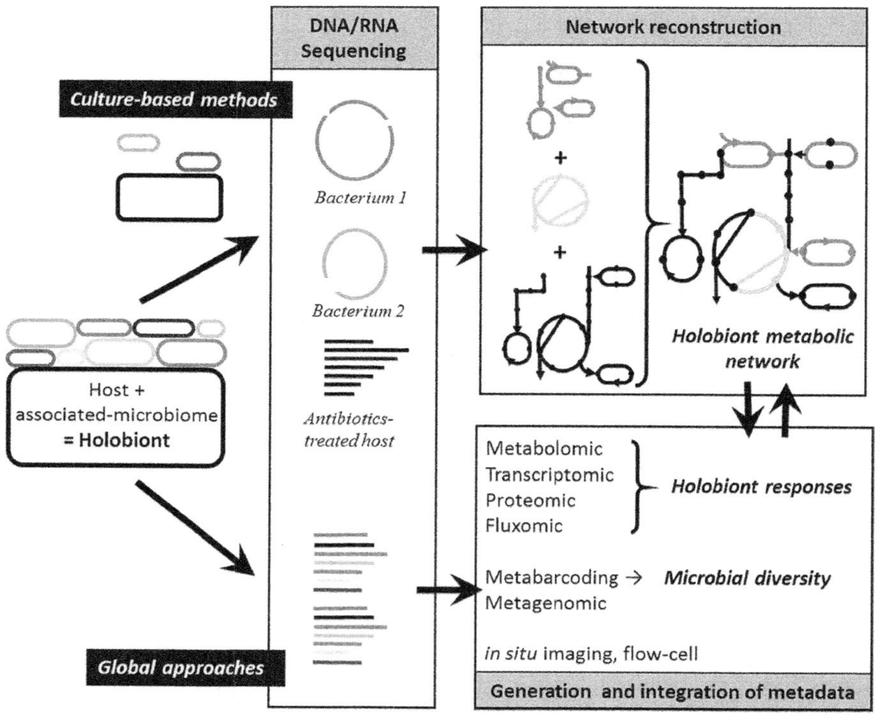

Figure 4.1. *Strategies for study of the microbiota and its host in chemical ecology (from [DIT 14])*

To validate the identified candidates (genes, enzymes, metabolites), the manipulation of the biological interaction in the laboratory constitutes a very efficient strategy when we can dispose of genetic systems, that is, either functional mutants (heterologous expression of the target gene and signal production in another bacteria) or "Knockout (KO)" mutants (invalidation by KO of the gene in associated bacteria and suppression of the signal production). If the identified chemical compounds are available, like certain phytohormones, their effect can be tested directly on the host in the absence of microbial partners.

In parallel with these targeted functional validation approaches, different imaging techniques can be used for studying the microbiota *in vivo* and the underlying chemical interactions (see also Chapter 7). For example, the bacterial culture on a flow-cell system helps visualize and quantify the growth kinetics of a biofilm by fluorescence and confocal microscopy in order to identify the molecules which contribute to chemical communication. The development of imaging techniques using MS (imaging by matrix-assisted laser desorption/Ionization (MALDI) or imaging by matrix-free uv-laser desorption/ionization (LDI)) offers the possibility of detecting and locating the signal metabolites on biological material sections at the tissular scale. Finally, other imaging approaches (nano-SIMS, secondary ion mass spectroscopy) combine *in situ* hybridization and isotopic labeling to identify bacteria at the genus and species level, and study their metabolism at the cellular scale.

4.3. The molecular dialog of microorganisms

To interact with one another and respond to the different environmental stimuli, microorganisms communicate with their own chemical language and use a wide range of extracellular signals and cellular responses, leading to beneficial or deleterious effects for one or the other of the partners with sometimes very precarious equilibria between these different effects.

4.3.1. *Language and social life of microorganisms*

It appears increasingly evident that bacteria are not isolated organisms living alone, but which communicate actively among themselves with the help of a chemical language using signaling molecules. It also becomes clear that bacteria live in communities, in particular in biofilms, which necessitate intra- and interspecies forms of communication. Quorum sensing (QS) is one of the regulation mechanisms by which bacteria detect and control their own cellular density in a bacterial population, and control the coordinated expression of certain bacterial genes via the production, detection and response to the accumulation of signalization molecules, named autoinducers, which they export in their environment [JAY 08]. When a critical concentration of these autoinducers is achieved, in response to an increase in population density, specific response systems are activated, triggering the induction of certain genes. These signal molecules activate complex adaptive responses, in particular intracellular receptors in gram-negative bacteria, two-component regulatory systems in gram-positive bacteria and membrane transporters in gram-negative or -positive bacteria (Figure 4.2).

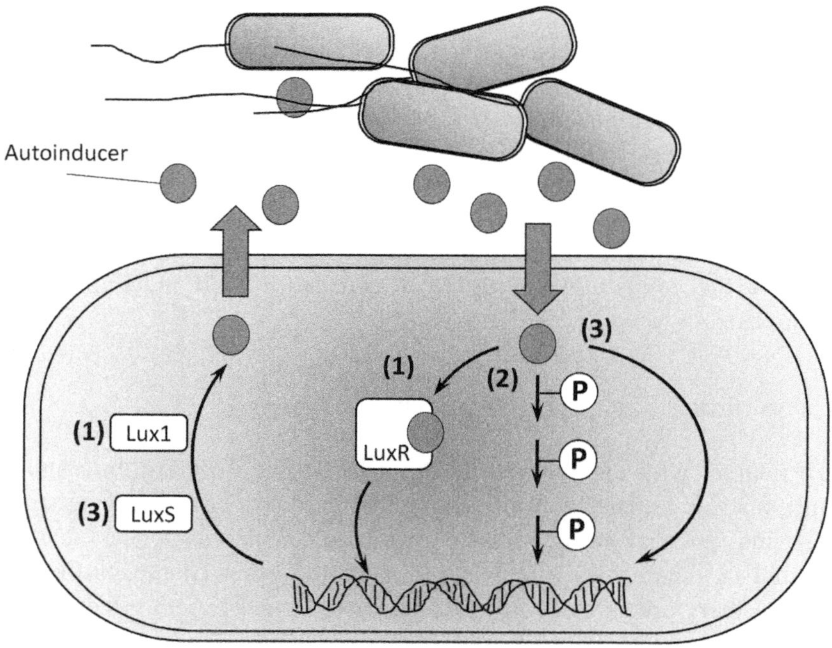

Figure 4.2. *Quorum sensing and its three principal signalization pathways: (1) the pathway involving intracellular receptors in gram-negative bacteria (LuxI/LuxR system) and homoserine lactones; (2) the two-component regulatory system in gram-positive bacteria involving peptidic autoinducers and a cascade of phosphorylation and (3) the pathway involving a membrane transporter (LuxS/Lsr) in gram-negative or -positive bacteria, and furanone-type autoinducers (from [RAF 05])*

These mechanisms regulate numerous important biological functions like virulence, sporulation, bioluminescence, horizontal gene transfer, biofilm formation and the production of various antibiotics and metabolites. QS was also studied in opportunistic pathogenic bacteria (*Staphylococcus aureus*, *Enterococcus faecalis*, *Streptococcus pneumoniae*, *Pseudomonas aeruginosa*, etc.) and has recently been proposed as a new target for antimicrobial therapy. Targeting the QS systems and their critical bacterial enzymes also constitutes a new strategy for the discovery of antibacterial agents less susceptible to generating resistances. Disturbing the cellular communication would also appear as an attractive alternative to avoid the formation of biofilms (see Chapter 9).

The autoinducers used by bacteria like communication molecules are essentially small molecules, like homoserine lactones in gram-negative bacteria (Figure 4.3), while gram-positive bacteria mainly use small peptides, such as cyclic peptides containing a thioester bond in *S. aureus*. The autoinducer role played by certain AMPs at sublethal concentrations has been identified, in particular in lactic acid bacteria. As such, their production depends on cellular density and their biosynthesis gene clusters encode proteins belonging to the family of two-component bacterial regulatory systems, including a histidine kinase playing the role of a sensor and a response regulator. Such autoregulatory circuits playing a key role in adaptation have already been identified for nisin (an AMP mainly used as a food preservative) and are actually highlighted for other classes of AMPs, such as siamycin or sviceucin (Figure 4.3). Thus, these autoinducer peptides including certain AMPs, which have the capacity to regulate bacterial growth by the intermediary of QS, open new tracks for the conception and development of antimicrobial molecules of high specificity without generating resistance [KLE 04].

More recently, QS has also been demonstrated in fungi. Thus, the pathogen *Candida albicans* is capable of producing farnesol, which inhibits the yeast-to-mycelium conversion, a differentiation process involved in the establisment of virulence. The role of farnesol was then generalized to ascomycetes. Meanwhile, despite profound genetic and molecular studies, the precise identification of the molecule governing the fungal QS processes (and fungal communication processes generally) still remains totally unknown, notably for the reason of their low production [LEE 11].

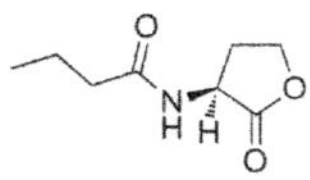

Example of acyl-homoserine lactone (C4-AHL)

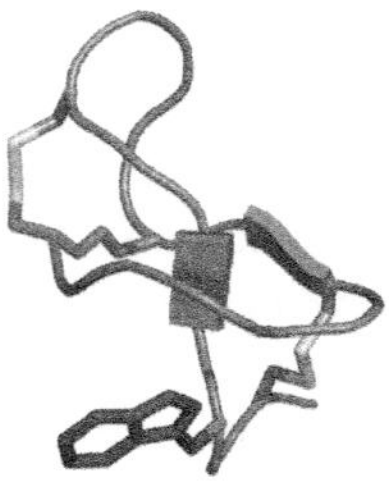

Antimicrobial Peptide Autoinducer
Three-dimensional structure of Sviceucin

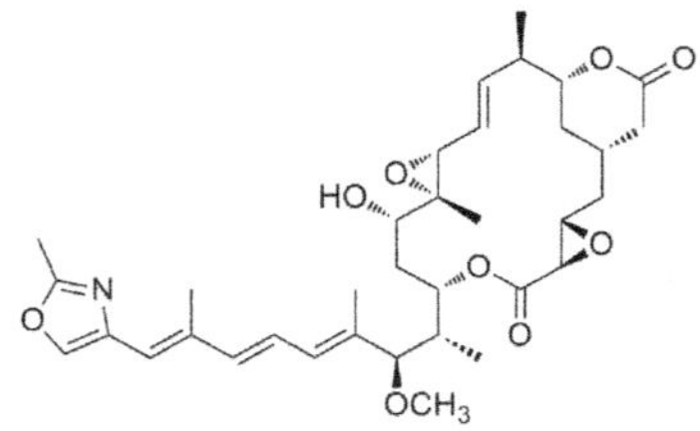

Rhizoxin

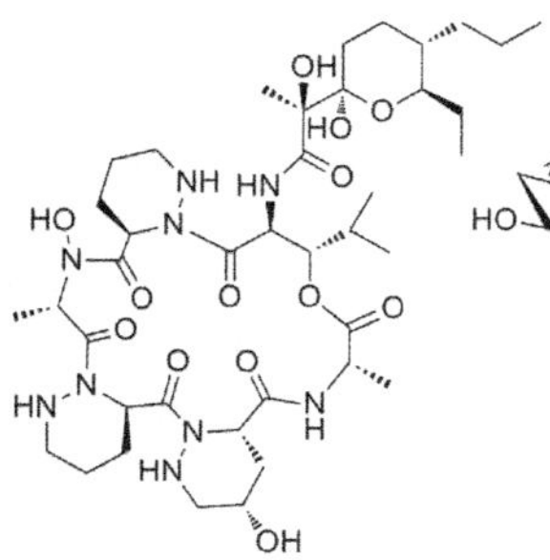

Dentigerumycin

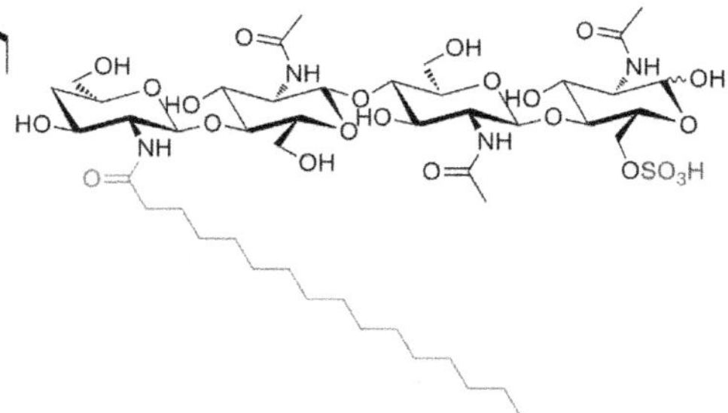

Chemical structure of one main Myc-LCO

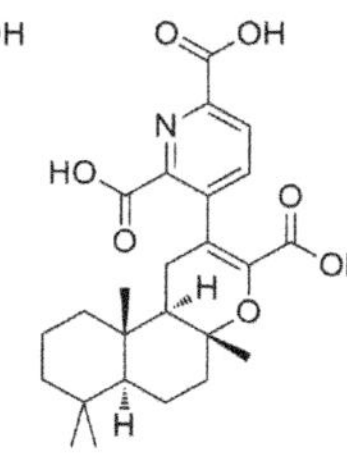

Thallusin

Figure 4.3. *Examples of semiochemicals involved in inter- or intramicrobiota interactions*

4.3.2. *The AMPs, main actors in the equilibrium of bacterial communities*

Diverse communities of microorganisms coexist and fight without mercy for nutrients and space, which are required for their growth and to overcome stress in a given biotope. In order to dominate an ecological niche, most bacteria are also provided with subtle and efficient mechanisms which help them to develop, to the detriment of others. Nutritional deficiency is one of the determining factors to trigger the production of an arsenal of molecules, and in particular antimicrobial peptides or proteins, which are key players of these mechanisms. Named bacteriocins in gram-positive bacteria, and microcins in gram-negative, these antimicrobial molecules are almost always directed against bacterial genera very close to the producing strain (narrow-spectrum antibiotics) [COT 13]. Meanwhile, it appears increasingly reasonable that they are not only a powerful defense and competition tool, but also a means of communication, as explained previously (see section 4.3.1). An important ecological consequence of the production of such peptides is the implementation and maintenance of microbial diversity in a given community. An elegant example of these interaction networks is described by the children's game rock-paper-scissors. According to the model, where the rock breaks the scissors, the scissors cut the paper and the paper envelops the rock, similarly a bacteriocin-producing bacterium kills a sensitive strain, but is supplanted by a strain which does not produce toxins and which is able to resist the toxin because it does not support the energetic cost of the toxin production. A resistant strain could itself be supplanted by a sensitive strain in absence of the toxin, the resistance also having a cost for the bacteria. It is in fact the impact of environmental pressures on the spatial structuring of the bacterial community, which does or does not help the coexistence of the three types of strains [MAJ 11]. Such interaction networks exist in all types of bacterial communities, whether they are housed by a host, like intestinal microbiota, or environmental.

4.3.3. *Fungi and bacteria communicate to better help each other*

Interactions between bacteria and fungi are widely present in the human organism, in food but also in nature. These interactions occur by physical associations between bacteria and planktonic fungi, by the creation of mixed biofilms or by the bacterial colonization of fungal hyphae. The resulting molecular associations between the two partners are by nature quite varied. Among this multitude of molecular communications, we can cite antibiosis (diffusion of molecules by a partner engendering a deleterious effect on the other), exchange of metabolites or genetic material, conversion of metabolites, secretion of proteins, signaling and chemotaxis. The range of these chemical interactions could have a direct impact on the physiology of the partner(s) and an inevitable incidence on their relationships as a consequence [FRE 11].

Among the positive associations between bacteria and fungi, symbiosis is the most widespread. This can be ectosymbiotic, the bacterium remaining at the exterior of the plasmic membrane of the fungus, or conversely endosymbiotic. In the context of the latter, metabolic cooperation was demonstrated between the fungal rice pathogen *Rhizopus microsporus* and the endosymbiotic bacterium *Burkholderia rhizoxinica* for the production of rhizoxin (Figure 4.3), a powerful anti-mitotic toxin which is crucial for the pathogenesis of the fungus. In fact, the joint analysis of the metabolome and the genome of the two partners showed the ability of the endosymbiotic bacterium to produce the main skeleton of the macrolide of the toxin, but harboring only one epoxide function, while two epoxide functions are naturally present in the molecule. This second function is actually introduced by a fungal oxygenase, helping to produce a molecule still more toxic for rice plants. This fantastic metabolic cooperation also contributes to the proof of concept of the evolution of natural products in symbiotic interactions.

4.3.4. *When helping each other degenerates into chemical warfare between bacteria and fungi*

Antibiosis is the best studied of the bacteria–fungi interactions and has led to the discovery of numerous antibiotics, the best-known example of which is penicillin, discovered following the observation of a strong antagonism between filamentous fungus *Penicillium* sp. and bacterium *Staphylococcus* sp. Recent examples have shown similar effects in the environment. Thus, certain *Pseudomonas* sp. can produce 2,4-diacetylphloroglucinol and phenazine types of compounds in the tomato, which induce the expression of ABC transporters in the phytopathogen fungus *Botrytis cinerea*, therefore limiting the accumulation of antifungal compounds in the hyphae. Indeed, laccase of the pathogen fungus is responsible for the production of reactive species capable of detoxifying the 2,4-diacetylphloroglucinol [SCH 02].

4.3.5. *The Trichoderma fungi: heavy artillery against pathogenic fungi*

Pathogenic fungi of the genus *Botrytis (B. cinerea, B. cacao)* are capable of infecting leaves, flowers and fruits of diverse cultures, for example, the grapevine, strawberry or cocoa bean. Therefore, they produce considerable damage in agriculture. Meanwhile, when the infection occurs at a favorable moment at the end of the season, *B. cinerea* also has beneficial effects: it is, therefore, named “noble rot” and leads to the elaboration of renowned sweet wines like Sauterne or Tokay.

Fungi of the genus *Trichoderma* (*T. harzianum*, *T. viride*, *T. longibrachiatum*, etc.) have a very rapid growth and exert a particularly effective antifungal activity against various pathogens, in particular against *Botrytis* sp., leading moreover to their use as biological control agents against fungal infections by *Botrytis* [VOS 14]. This biocontrol performed by the *Trichoderma* is explained by many mechanisms and relies on the production of a powerful and varied chemical arsenal, leading to the establishment of an elaborate battle strategy. The first line of attack is afforded by the production of metabolites, which are small volatile molecules (e.g. *pentyl-pyrones*). Then, the heavy artillery is implemented with the secretion of antifungal peptides called peptaibols. These peptides rapidly kill the other fungi by perturbing the stability and permeability of their plasmic membranes in which they form pores, resulting in the escape of vital cellular compounds and the death of the fungi. Finally, the last line of attack is supplied by cellulolytic enzymes. This triple strategy, already very effective as is, is reinforced by a synergy between the peptaibols and the hydrolytic enzymes of the cell walls. By their capacity to perturb membranes, peptaibols in fact inhibit the membrane-associated enzymes, which ensure synthesis of the glycans, which are constituents of the cell walls. They thus inhibit the re-synthesis of the beta-glycans of the cell wall and maintain the activity of glucanases, which destroy the membrane. Reconstruction of the cell walls is thus a major target of the antagonistic activity developed by *Trichoderma.* The combined action of these various molecules thus diversifies the mechanisms leading to the antifungal activity, which finds itself amplified.

4.4. Chemical communication between microorganisms and their hosts

Whether they are plants, marine invertebrates or insects, all have acquired complex association with microorganisms in the course of evolution, which are often beneficial. Even if the deciphering of these interactions is not yet completely achieved, current knowledge already testifies to the preponderant role of chemical signaling in the establishment of these associations.

4.4.1. *Plant–bacteria relationships: essential interactions with different partners*

Some plants have acquired the capacity to form symbiotic associations with nitrogen-fixing bacteria to take advantage of the nitrogen reservoir present in the atmosphere (in the form N_2). The associations lead to the formation of new root organs, called nodules, in which the bacteria live on the plant cells. This allows an effective exchange of nutrients between the two symbiotic partners. Two groups of plants can form nitrogen-fixing nodules: the leguminous plants with bacteria of the genus *Rhizobium* and the actinorhizal plants, which associate with the actinomycetes *Frankia* [OLD 13].

4.4.1.1. *Symbiotic relationships in the rhizosphere between leguminous plants and rhizobium*

An intimate symbiotic type of relationship exists between leguminous plants and bacteria of the genus *Rhizobium* (called "rhizobium") in the root system of the plant. The first step of this interaction consists of specific molecular recognition of two partners, involving an interaction between the bacterial receptors of the external envelope and molecules of plant origin, the flavonoids. The fixation of flavonoids on the receptors induces the expression of a specific nodulation gene in bacteria, called *nodD*, which encodes the inactive protein NodD. The protein NodD could activate the other *nodD* genes, which are normally silent and the products of which are involved in the biosynthesis of particular oligosaccharides called Nod factors. These factors are specific to each bacterial type that the plant can in turn detect. The following step is the formation of root nodules, where the organic molecules necessary for the mutual food needs of both partners can be exchanged. In these nodules, the reduction of the atmospheric nitrogen in ammonium, which is a natural fertilizer, takes place. In turn, the bacteria use the carbon substrates produced by plant photosynthesis (Figure 4.4).

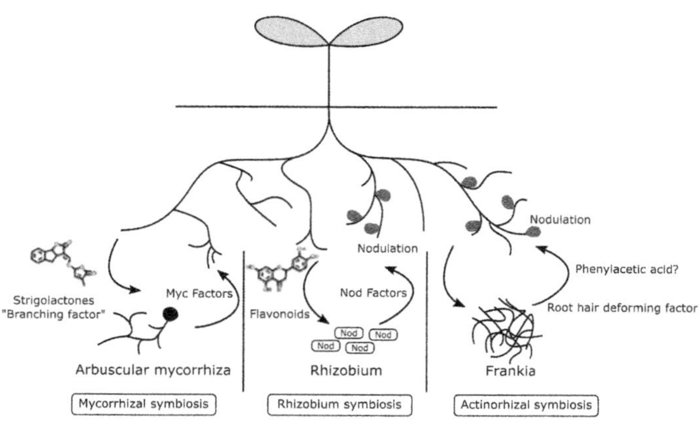

Figure 4.4. *Molecular dialog between the plant and soil microorganisms for the establishment of symbiotic interactions (from [ABD 12])*

4.4.1.2. *Actinorhizal symbiosis: another symbiotic relation of the rhizosphere*

Bacteria of the genus *Frankia* are actinomycetes, which live in symbiosis on the rhizosphere of the actinorhizal plants (250 known species of plants spread among 8 families and 25 genera), thus allowing them to fix atmospheric nitrogen, by reducing N_2, and adapt to the conditions of deficiency. Meanwhile the molecular dialog between the plant and the associated actinomycetes remains poorly understood. It has nonetheless been demonstrated that *Frankia* will produce a "root hair deforming factor", a thermostable compound of size less than 3 kDa, sensitive to certain enzymes (chitinase), which will be involved in the deformation of the root hairs of the plant as a prelude to colonization. Likewise, another signal molecule of *Frankia* is phenylacetate (PAA), an aromatic acid analogous to auxin, a plant hormone. This compound will induce the synthesis of adventitious secondary roots, which will then be colonized by the bacteria. Nonetheless, the chemical nature of the molecules produced by the plant for establishment of the symbiosis remains yet unknown (Figure 4.4).

4.4.2. *Plants also establish intimate relations with fungi*

All plants live in more or less prolonged association with fungi. They may live as symbionts between the plant cells and develop either in a harmonious manner (endophytes), or on the contrary provoke illness (phytopathogens). Mycorrhizal fungi, meanwhile, are among mutualistic symbionts the most important of the rhizosphere. They form particular structures with the roots of plants, mycorrhizae, which participate in the nutrition of almost all plants.

4.4.2.1. *Mycorrhizal symbiosis: another chemical interaction model beneficial and essential to the rhizosphere*

Mycorrhizal fungi are incapable of photosynthesis and are therefore completely dependent for their carbonaceous substances on the plant which they colonize. In turn, they supply nitrogen, phosphorus and other mineral substances which they are capable of mobilizing from the connection of their hyphae with the soil.

Mycorrhizae can be morphologically and phylogenetically classified into two main groups:

– ectomycorrhizae, where the fungi do not penetrate inside the root cells. They form at the exterior of the roots a sort of fungal mantle. Among the ectomycorrhizae, arbuscular ectomycorrhizae constitute the most studied group;

– endomycorrhizae, where fungi return to the interior of the root cells to carry out exchanges with the plant.

Arbuscular ectomycorrhizae symbiosis is established in a manner nearly identical to that of rhizobial symbiosis, through bilateral chemical signaling. The fungi are attracted by the semiochemicals produced by the plant: the strigolactones or "branching factor", the chemical nature of which has for a long time been unknown. These signals will then induce the expression of *myc* genes leading to the production of Myc (FMyc) factors, the structure of which is close to those of the NOD factors. A cascade of symbiotic signals follows, leading to the intercellular colonization of the fungal hyphae in the root cortex of the plant and the formation of arbuscules (Figure 4.4).

4.4.2.2. *The important assistance of fungal endophytes*

Most plants are colonized by fungi without developing symptoms of disease. These microorganisms, qualified as "endophytes", invade the internal tissues of living plants (leaf, stem, root) and maintain a mutualistic relationship with the plant host. We distinguish two large groups of endophytes with very different lifestyles: those which infect herbaceous plants (*Clavicipitaceae*) and those, more diverse, not belonging to this family (non-*Clavicipitaceae*).

Some of these endophytes are able to reinforce the tolerance of the host plant not only to environmental stresses (dryness, large temperature variations, high salinity, etc.), but also to attacks from herbivores and phytopathogenic fungi. For example, the presence of endophytes in the leaves of *Theobroma cacao* (cocoa) significantly reduces necrosis and mortality of leaves when the plant is infested by the oomycete *Phytophthora* sp.

The exact physiological protection mechanisms of the plants by the endophytes are still unknown. Meanwhile, many studies increasingly attest that endophytic fungi would confer to the host plants a certain number of advantages linked to the production of chemical compounds, and that, in turn, they would benefit from the protection and nutrients of the plant.

Numerous compounds produced by endophytic fungi have demonstrated fungicide, herbicide or antibacterial properties, reinforcing the idea that these molecules play an important role in the defense of the host plant. It has been proven, for example, that the significant antifungal activity of the corn endophyte *Acremonium zeae* against the fungal pathogenic fungi *Aspergillus flavus* and *Fusarium verticillioides* is linked to the production of complex polyketides [WIC 09]. Similarly, certain endophytes are capable of producing toxins against herbivores, and the peramine produced by the endophyte fungus *Neotyphodium lolii* is a wonderful example of a molecule of fungal origin with anti-feedant properties for insects.

4.4.2.3. *When fungi become enemies of plants*

Among the interactions between fungi and plants, parasitism is highly represented and constitutes one of the main damages for agricultural production in France and Europe, in particular for large cultures (cereals, oilseeds, grapes). Fungi use various strategies to infect plants. Meanwhile, in most cases, plant infection by fungi is correlated with the production of potent toxins with deleterious effects for the plant.

In reaction to this infection, the plant produces defense molecules, which are either constitutive (phytoanticipins) or *de novo* induced in reaction to the attack (phytoalexins). Nonetheless, pathogenic fungus pathogens have developed detoxification processes of these defense secondary metabolites, which are mediated by their chemical transformation into less toxic compounds.

4.4.3. *Mutualist actinobacteria provide care to insects*

In insects, mutualist microorganisms are well known for their capacity to contribute to nutrition or to their host's reproduction. Their role is also important for the protection of the host against attacks by pathogens. An important number of gram-positive bacteria belonging to the actinobacteria group, in particular of the genus *Streptomyces*, contribute to the defense and survival of various insects, ants, beetles or wasps, by the production of antimicrobials. Symbionts can be transmitted vertically through the intermediary of eggs, or horizontally by coprophagy. The role of actinobacteria varies, from protection of food sources to the protection of the insects themselves against their pathogens.

Certain ants of the genus *Atta*, called "leaf-cutter", are "cultivator" ants, living mainly in the Amazon, in the humid forests of the United States or semi-arid regions (Mexico, Arizona, etc.). They have an ecologically and economically important role, creating problems in agriculture and agroforestry, because of their extensive colonies that sometimes cause significant damage. They very efficiently cut the hardest leaves thanks to their very powerful and sharp mandibles, thus defoliating fruit trees and crops. They are not nourished by these leaves, but use them as support for the cultivation of fungi on which they and their larvae feed [WEB 66]. They in fact practice mono-culture by selecting Ascomycetes of the genus *Apterostigma*. Actinobacteria (*Pseudonocardia*, *Amycolatopsis*) hosted by the insect are, therefore, capable of secreting antimicrobials, like dentigerumycin (Figure 4.3), which serve as "food preservers" and thus prevent the food fungi of *Atta* ants from being destroyed or damaged by infections and alterations provoked by other fungal parasites (Ascomycetes of the genus *Escovopsis*). The *Pseudonocardia* symbiotes are essentially transmitted vertically via the queens, who found the ant colonies, but cases of horizontal transmission or *de novo* acquisition from the environment are encountered, testifying to a strong dynamic of this association. The symbiotic associations with actinomycete bacteria displayed in the leaf-cutter ants, or in the pine beetle, suggest that these insects establish highly dynamic associations for the protection of their food reserves against fungi which spoil them, by the acquisition of symbionts from their environment and the production of compounds which are beneficial to them.

Symbiotic interactions with bacteria, in particular the actinobacteria, are equally essential for the protection of insects against pathogens, or predators, and for their reproduction [KUT 15, SEI 12]. A very good example is that of the highly specific association between the actinobacteria *Candidatus* Streptomyces philanthi and solitary bee-eating wasps of the genus *Philanthus*, which attack and capture other Hymenoptera – such as bees – to eat their larvae. These bacteria protect the offspring of the wasp from infections by pathogens. The female bee-eating wasps cultivate the bacteria *Candidatus* Streptomyces philanthi in the glands of their antennae and deposit them on their nest before laying. The larvae then transfer the symbiotic bacteria to their cocoon, where they provide a large and effective protection against bacteria and pathogen fungi by the production of antimicrobial molecules (so far unidentified). Females probably acquire the symbiont at the time of their emergence out of the cocoon, by exploration and contact with its surface [KAL 06].

It is likely that the resistance of actinobacteria to unfavorable conditions, and their ability to form spores and to synthesize a large number of metabolites displaying usually antibacterial or antifungal properties, are at the origin of their evolutionary success as insect symbionts. Insects having engaged in symbiotic interactions with these antibiotic-producing bacteria have thus found effective partners, as much for their protection and those of their descendants against pathogens and competitors, as for successful conservation of their food resources [KAL 09].

4.4.4. *Chemical communication between microorganisms and their host in the marine environment*

4.4.4.1. *Deciphering a beneficial chemical interaction between algae and bacteria*

Marine algae grown in laboratory culture conditions have shown that certain species are not capable of survival in axenic conditions, that is, in the absence of bacteria, and thus present abnormal development. Some green algae of the order Ulvales present filamentous forms in axenic conditions, while they develop in the form of foliated thalli in the presence of bacteria. Thallusin (Figure 4.3), isolated from a bacterium belonging to the Cytophaga–Flavobacteria–Bacteroidetes (CFB) group, is the compound responsible for this effect on the morphology of Ulvae, and also plays a role in the germination of spores [MAT 05]. The microbiota associated with marine macroalgae is quite variable not only from one species to another, but also along the thallus (Figure 4.5). Certain dominant bacterial groups, such as the gamma-proteobacteria or the CFB, synthesize numerous chemical compounds, potentially morphogens, but for which the signal roles remain to be explored in marine algae. On the contrary, it is well-known that the red alga *Delisea pulchra* produces halogenated furanone which is capable of inhibiting the QS of pathogen bacteria involved in the bleaching of the algal host [HAR 12].

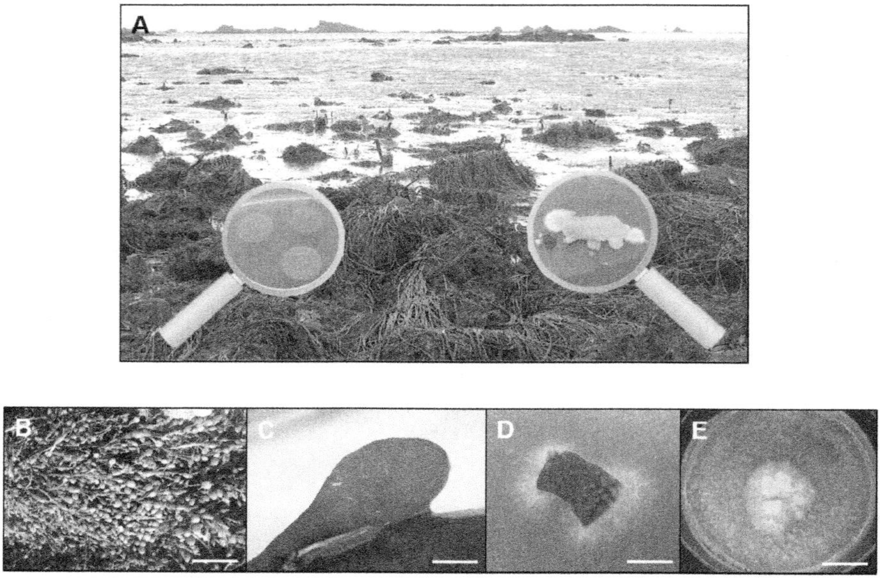

Figure 4.5. *Illustrations of the fungal and bacterial endomicrobiota associated with laminaria. B-E) Examples of cultivable endophytic fungus associated with Ascophyllum nodosum. B) Ascophyllum nodosum (scale 10 cm), C) receptacle of Ascophyllum nodosum (scale: 1 cm), D) isolation of the endophytic marine ascomycete AN129 R from the receptacles of A. nodosum (scale: 1 cm), E) culture of the AN 129 R isolate (scale: 1.5 cm). (copyright: Marine Vallet)*

In marine environments, chemical communication seems also to be at the center of the dynamic of certain planktonic populations, but the nature of the chemical signals and the mechanisms responsible for their biosynthesis remain unknown [MAR 14, WAH 12].

4.4.4.2. *A dense and complex chemical language between marine invertebrates and their associated bacteria*

The sedentary lifestyle of benthic sessile invertebrates, such as sponges and corals, makes them particularly vulnerable to predation, competition and environmental changes (see Chapters 1 and 2). These metazoans have developed different defense strategies, including the production of defense and communication molecules, and also the capacity to host and favor an association which is more or less rich and more or less stable with bacteria which produce molecules from which they benefit. The sponge and its associated microbiota respond to the definition of the holobiont, which in fact constitutes a metaorganism. In the holobiont, chemical communication between the cells of the sponge host, its numerous microbial partners and the environment is essential for the survival of the host and its associated microorganisms. These bacterial associations could represent up to 40% of the biomass of the sponge in certain species. Meanwhile, the mechanisms by which these complex associations have been put in place in the course of evolution and are maintained are still very poorly understood: simultaneous acquisitions by vertical transmission up to different stages of development and horizontal transmission by filtration activity have been proposed. Bacteria find a biotope in the host where they benefit from efficient protection and, at the same time, a very rich nutrient and oxygen supply by the circulation of water in internal cavities. From its side, the host depends on bacteria and eukaryotic microorganisms for its nutrition, by filtration or by a stable culture of certain microorganisms. A complex network of signaling molecules, within the bacterial populations and communities and with their host sponge, is established in this manner. These molecules participate in QS, are involved in bacterial competition in the microbiota and participate in the chemical defense of the sponge or help it to colonize a biotope, via the production of several molecules. Certain compounds help the fixation of larvae and their metamorphosis, the lifecycle of sponges and corals involving the alternation of mobile larvae and fixed adults. The larval metamorphosis of certain scleractinian coral like *Pocillopora damicornis* or the demosponge *Rhopaloeides odorabile* is induced by biofilm formation on their surface. Recent studies have shown that this process is mediated by chemical compounds of yet unknown structures. Photoprotective or antioxidant molecules (carotenoids and other pigments), produced by microorganisms, are also used as photoprotectors, the sponges and coral which live in coastal zones being regularly subjected to intense solar radiation causing UV stress [QUÉ 14].

4.5. Regulations and evolution of the interactions in changing ecosystems and environments

Intra- or intermicrobiota chemical interactions are subject to numerous variations and regulations, which lead to notable modifications in the production of semiochemicals which we have started to understand, notably from a better knowledge of genomes and the mechanisms involved in the biosynthesis of these compounds.

4.5.1. *Contribution of chemical ecology to the understanding of biosynthesis mechanisms of chemical mediators*

Semiochemicals produced by microorganisms and involved in their biological and ecological functions result from the succession of enzymatic reactions starting from very simple chemical precursors. The enzymes involved in these biosynthesis pathways are also able, from an elementary brick, to construct elaborate molecular buildings. These biosynthesis enzymes (e.g. polyketide synthases (PKS), non-ribosomal peptide synthetases (NRPS) or terpene synthases) are encoded by genes that are generally organized in clusters in bacteria and fungi. They are otherwise often assisted by auxiliary enzymes (tailoring enzymes) able to perform supplementary decorations on the general skeleton generated during biosynthesis.

Meanwhile, the growing access to entire genomes of microorganisms, associated with the progress of bioinformatics analysis, has clearly demonstrated that the number of biosynthesis gene clusters was much larger than initially predicted from identified metabolites. However, a large number of "cryptic" or "silent" gene clusters have been identified, which cannot be expressed in laboratory conditions and whose biosynthesis pathway and the metabolite(s) generated are unknown. The expression of these "silent" clusters can be controlled by a large network of regulations involving numerous enzymes, themselves under the control of environmental stimuli such as pH, temperature, light, the formation of biofilms, and also the communication between microorganisms.

Thus, it has been recently demonstrated that the physical interaction between the *Aspergillus nidulans* fungus and the *Streptomyces rapamycinicus* soil bacteria resulted in the activation of a PKS "cryptic" fungal gene, helping the biosynthesis of orsellinic acid and its derivatives. The precise mechanism of this interaction otherwise demonstrated that the activation of the PKS fungal gene was linked to the acetylation of fungal histones catalyzed by a Histone Acetyl Transferase (HAT) from the bacteria [BRA 13].

It is clear that chemical ecology, through the study of the interactions between microorganisms, contributes to knowledge not only for the research of new secondary metabolites and the chemical understanding of their biosynthesis pathways, but also for the study of the genomic regulation of cryptic gene clusters. Otherwise, the range of these data suggests that the current microorganism culture techniques in view of the isolation of chemical mediators must be rethought, and the co-culture techniques to mimic the microbiota appear to be a promising approach. Numerous fruitful examples in the literature have validated this approach for the activation of silent genes and the production of original molecules.

4.5.2. *Metabolic networks: new tools for studying the evolution of host/microbiota interactions*

Environmental changes, whether they are punctual (short-term abiotic stress) or of a longer duration, result in functional and metabolic variations affecting both the host and its associated microbiota. This can lead to deep modifications of the interactions, such as disturbance of the molecular dialog, instability of the symbiosis, change in the surface microbial community, emergence of pathogens, etc. In the long term, new equilibria appear together with adaptive phenomena of the holobiont. In this context, the association of metabolomics with genomics and post-genomics analyses tools is a real asset for exploring these co-evolutionary processes. The study of components of the holobiont at different levels (genomics, transcriptomics, metabolomics; see Chapters 6 and 7) and the reconstruction of molecular networks could in fact help to study in a global manner the behavior of the integrated metabolic network and the acclimation processes of the holobiont in response to environmental disturbances (Figure 4.1), [DIT 14].

4.6. Conclusions – from chemical ecology to future applications: impacts of the study of the microbiota

Numerous examples of inter- and intramicrobiota interactions cited in this chapter attest to their generalized presence in numerous ecosystems and their fundamental impact, notably at the environmental process level. Deciphering these interactions in the biomolecular continuum (from the gene to the molecule) is, therefore, a prerequisite to be able to better understand them from a fundamental point of view, and also to better control them and/or evaluate them (see also Chapter 9).

The discovery during the last 30 years of an extraordinary diversity of microorganisms, inhabiting environments which we have previously believed to be hostile to life, also opened new perspectives for the study of functionality and stability of their biological macromolecules, as well as the origins of life on Earth and on the exoplanets.

These "extremophile" microorganisms and Archaea, which constitute the 3rd domain of life (the three domains of life being Bacteria, Archaea and Eukarya), also use QS. Thus, the study of QS in these microorganisms is one of the expanding domains of research because of their biotechnological effects, notably for the use of extracellular enzymes or metabolites, but also for knowledge of the mechanisms of adaptation of microorganisms.

In addition, the advent of new spectroscopic (high field NMR, see Chapter 8) and analytical (metabolomic, see Chapter 7) techniques and genomics (see Chapter 6) associated with the expansion of *in situ* imaging techniques and global methods for studying the holobiont (metabolic networks) must help to better track these mediators, which are often produced in very small quantities, in specific conditions and are subject to numerous variations. Likewise, by witnessing recent advances in genomics, a large number of chemical mediators involved in microbial communications do not express in standard laboratory conditions, suggesting a higher chemical diversity than what has already been discovered. The characterization of these new compounds and determination of their ecological roles should, therefore, help the emergence of new compounds of high added value in domains such as medicine, ecology, agronomy or even the biotechnologies.

4.7. Bibliography

[ABD 12] ABDEL-LATEIF K., BOGUSZ D., HOCHER V., "The role of flavonoids in the establishment of plant roots endosymbioses with arbuscular mycorrhiza fungi, rhizobia and *Frankia* bacteria", *Plant Signaling & Behavior*, vol. 7, no. 6, pp. 636–641, 2012.

[BRA 13] BRAKHAGE A.A., "Regulation of fungal secondary metabolism", *Nature Reviews Microbiology*, vol. 11, pp. 21–32, 2013.

[COT 13] COTTER P.D., ROSS R.P., HILL C., "Bacteriocins – a viable alternative to antibiotics?", *Nature Reviews Microbiology*, vol. 11, pp. 95–105, 2013.

[DIT 14] DITTAMI S.M., EVEILLARD D., TONON T., "A metabolic approach to study algal–bacterial interactions in changing environments", *Molecular Ecology*, vol. 23, pp. 1656–1660, 2014.

[FRE 11] FREY-KLETT P., BURLINSON P., DEVEAU A. *et al.*, "Bacterial-fungal interactions: hyphens between agricultural, clinical, environmental, and food microbiologists", *Microbiology and Molecular Biology Reviews*, vol. 75, pp. 583–609, 2011.

[HAR 12] HARDER T., CAMPBELL A.H., EGAN S. *et al.*, "Chemical mediation of ternary interactions between marine holobionts and their environment as exemplified by the red alga *Deliseapulchra*", *Journal of Chemical Ecology*, vol. 38, pp. 442–450, 2012.

[JAY 08] JAYARAMAN A., "Bacterial quorum sensing: signals, circuits and implications for biofilms and disease", *Annual Review of Biomedical Engineering*, vol. 10, pp. 145–167, 2008.

[KAL 06] KALTENPOTH M., GOETTLER W., DALE C. *et al.*, "'*Candidatus* Streptomyces philanthi', an endosymbiotic streptomycete in the antennae of *Philanthus* digger wasps", *International Journal of Systematic and Evolutionary Microbiology*, vol. 56, pp. 1403–1411, 2006.

[KAL 09] KALTENPOTH M., "Actinobacteria as mutualists: general healthcare for insects", *Trends in Microbiology*, vol. 17, pp. 529–535, 2009.

[KLE 04] KLEEREBEZEM M., "Quorum sensing control of lantibiotic production: nisin and subtilinautoregulate their own biosynthesis", *Peptides* vol. 25, pp. 1405–1414, 2004.

[KUR 15] KURTBÖKE D.I., FRENCH J.R., HAYES R.A. *et al.*, "Eco-taxonomic insights into actinomycete symbionts of termites for discovery of novel bioactive compounds", *Advances in Biochemical Engineering/Biotechnology*, vol. 147, pp. 111–135, 2015.

[LEE 11] LEEDER A.C., PALMA-GUERRERO J., GLASS N.L., "The social network: deciphering fungal language", *Nature Reviews Microbiology*, vol. 9, pp. 440–451, 2011.

[MAJ 11] MAJEED H., GILLOR O., KERR B. *et al.*, "Competitive interactions in *Escherichia coli* populations: the role of bacteriocins", *The ISME Journal*, vol. 5, pp. 71–81, 2011.

[MAR 14] MARTIN M., PORTETELLE D., MICHEL G. *et al.*, "Microorganisms living on macroalgae: diversity, interactions, and biotechnological applications", *Applied Microbiology and Biotechnology*, vol. 98, pp. 2917–2935, 2014.

[MAT 05] MATSUO Y., IMAGAWA H., NISHIZAWA M. *et al.*, "Isolation of an algal morphogenesis inducer from a marine bacterium", *Science*, vol. 5715, p. 1598, 2005.

[OLD 13] OLDROYD G.E.D., "Speak, friend, and enter: signalling systems that promote beneficial symbiotic associations in plants", *Nature Reviews Microbiology*, vol. 11, pp. 252–263, 2013.

[QUÉ 14] QUÉVRAIN E., DOMART-COULON I., BOURGUET-KONDRACKI M.-L., "Marine natural products: chemical defense/chemical communication in sponges and corals", in OSBOURNE A., GOSS R.J., CARTER G.T. (eds), *Natural Products: Discourse, Diversity and Design*, John Wiley & Sons, 2014.

[RAF 05] RAFFA R.B., IANNUZZO J.R., LEVINE D.R. *et al.*, "Bacterial communication ("Quorum Sensing") via ligands and receptors: A novel pharmacologic target for the design of antibiotic drugs", *The Journal of Pharmacology and Experimental Therapeutics*, vol. 312, pp. 417–423, 2005.

[SCH 02] SCHOONBEEK H.J.1., RAAIJMAKERS J.M., DE WAARD M.A., "Fungal ABC transporters and microbial interactions in natural environments", *Molecular Plant-Microbe Interaction*, vol. 15, pp. 1165–1172, 2002.

[SEI 12] SEIPKE R.F., KALTENPOTH M., HUTCHINGS M.I., "*Streptomyces* as symbionts: an emerging and widespread theme?", *FEMS Microbiology Reviews*, vol. 36, pp. 862–876, 2012.

[VOS 14] VOS C.M., DE CREMER K., CAMMUE B.P. *et al.*, "The toolbox of *Trichoderma* spp. in the biocontrol of *Botrytis cinerea* disease", *Molecular Plant Pathology*, vol. 16, pp. 400–412, 2014.

[WAH 12] WAHL M., GOECKE F., LABES A. *et al.*, "The second skin: ecological role of epibiotic biofilms on marine organisms", *Frontiers in Microbiology*, vol. 3, p. 292, 2012.

[WEB 66] WEBER N.A., "Fungus-growing ants", *Science*, vol. 153, pp. 587–604, 1966.

[WIC 09] WICKLOW D.T., POLING S.M., "Antimicrobial activity of pyrrocidines from Acremoniumzeae against endophytes and pathogens of maize", *Phytopathology*, vol. 99, no. 1, pp. 109–115, 2009.

5

From Chemical Ecology to Ecogeochemistry

The entire biosphere is today subjected to the consequences of human activities and to their combined impacts on the environment and biodiversity. These impacts often translate into modifications of the chemical properties of the environment which are notably observed in the most remote areas of the globe (contamination by organic pollutants persisting in polar regions, acidification and anoxia of deep water of certain oceanic regions, eutrophication and/or soil erosion). Despite this, the concepts and tools to understand the impacts of these multiple abiotic constraints at different spatio-temporal scales, on diversified communities and in interactions are missing.

If the tolerance of species or the metabolic responses of organisms to stress are an indication of the role these constraints play in the distribution of species and population structures, their influence on biodiversity dynamics and ecosystem functioning depends, on the contrary, on reciprocal relationships between organisms and abiotic components in their environment which are not yet well understood. This chapter illustrates the complexity of interactions between biotic and abiotic ecosystem components, and shows how different types of interactions are established by direct modification of their immediate chemical environment by the communities of organisms, via active biological and chemical components, generating feedback loops that can, therefore, be qualified as ecogeochemicals.

Chapter written by Catherine FERNANDEZ, Virginie BALDY and Nadine LE BRIS.

5.1. Balance between primary and secondary metabolism

Sessile (e.g. plants, macrophytes or benthic animals) but also free (e.g. phytoplankton) organisms produce primary metabolites necessary for their growth and reproduction. The number of these primary metabolites is probably below 10,000 and involves 50% of the genes of a species [PIC 11]. The organisms also produce a myriad of specialized secondary metabolites necessary for their survival and to interaction with their environment. The number of secondary metabolites is estimated to be approximately 200,000 (but some authors suggest that this number is probably largely underestimated) and brings into play 10–20% of organism genes. These compounds can be roughly divided into three principal classes, namely phenolic compounds, terpenoids and alkaloids (Figure 5.1 [HOP 03]) (see also Chapter 2).

The phenolic compounds are characterized by the presence of at least one hydroxyl aromatic ring and are hydrophilic. They are highly diverse, ranging from simple phenols, phenolic acids and flavonoids to large complex polymers such as tannins and lignins. They have various functions in plants: in the structure of the cellular wall (lignin), in providing color, fragrance and taste (flavonoids), protecting plants against damages linked to UV, herbivores, insects and microorganisms, and in biotic interactions as allelochemicals.

Terpenes are generally lipophilic compounds based on a simple five-carbon building block (isoprene unit). Monoterpenes are composed of two isoprene units (C10), sesquiterpenes possess three isoprene units (C15), whereas diterpenes have four (C20), etc. Some terpenes, mainly isoprenes, mono- and sesquiterpenes are volatile and are referred to as volatile organic compounds (VOC). The terpene family includes hormones, carotenoid pigments, latex and most essential oils [HOP 03]. Terpenes play different roles such as attracting pollinators and protecting against herbivores, toxic insecticides and insect repellents [HOP 03].

Alkaloids are soluble compounds characterized by at least one nitrogen atom and have a high biological activity, like analgesic or toxic properties [HOP 03]. At higher doses, most alkaloids are very toxic (morphine, nicotine, caffeine, quinine, etc.). These compounds have been mainly studied in pharmacology and used as drugs or medication. We can cite, for example, quinine, an analgesic and anti-malarial compound.

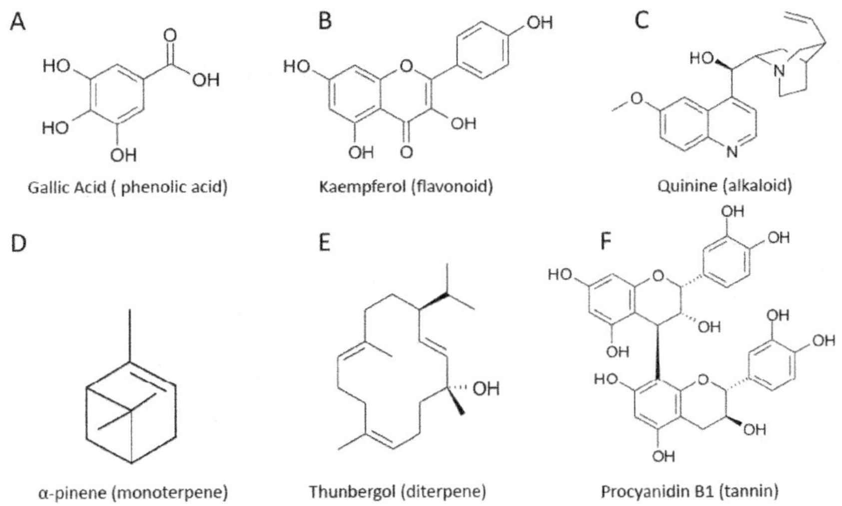

Figure 5.1. *Examples of different types of secondary metabolites (A: Phenolic acids, B: Flavonoids, C: Alkaloids, D: Monoterpenes, E: Diterpenes, F: Tannins)*

The multiplicity of functions of secondary compounds is very important as concluded by Williams *et al.* [WIL 89], in their review on these functions, that "the secondary metabolites are a measure of the capacity of organisms to survive. The possibility of synthesizing a large range of secondary products, which can repel other organisms, appears to be one facet of the strategy of organisms for their survival". However, the defense function, one of the first studied, is considered primordial. Indeed, these organisms (plants or benthic organisms) being sessile, cannot escape from predators or pathogens' attacks or variations in environmental conditions. However, these organisms are not passive victims of these aggressions, since they can defend themselves by the production of these secondary compounds. These defenses can be constitutive (i.e. produced continuously by the organism) or induced (i.e. produced punctually in case of aggression). The production of these secondary products will, therefore, depend on the pressures on the organisms induced by biotic stress such as herbivory, pathogens, etc., or by abiotic stress such as temperature changes, pollution and also nutrients availability. The secondary compounds present large interpopulational variabilities and large phenotypic plasticity, which appears to be an adaptive trait contributing to their fitness. One of the hypotheses concerning the phenotypic variations of secondary compounds concerns the limitation of resources often present in the natural environment. Numerous studies have focused on the effect of soil nutrients on the production of secondary metabolites. The nutritional status of plants will alter the availability of precursors the synthesis of defenses and also change the plant physiology and the resource allocation pattern. Terpenoids, nitrogen and phosphorous in the soil influence production and emission of isoprene and/or monoterpenes not only through photosynthesis and production of precursors such as IPP [ORM 12], but also through parameters linked to cell anatomy (Figure 5.2). Generally, researchers have observed a positive relationship between nutrient content, such as nitrogen, and isoprene emission, but the effect on terpenoids' storage is more controversial and can be more contrasted according to the species considered. The impact of soil nutrient content on terpenoids is not generally observable except the quantities of compounds produced, but little on their diversity. Likewise, for phenolic compounds, the production varies as a function of nutrient availability and is not generally observed except for the quantities produced.

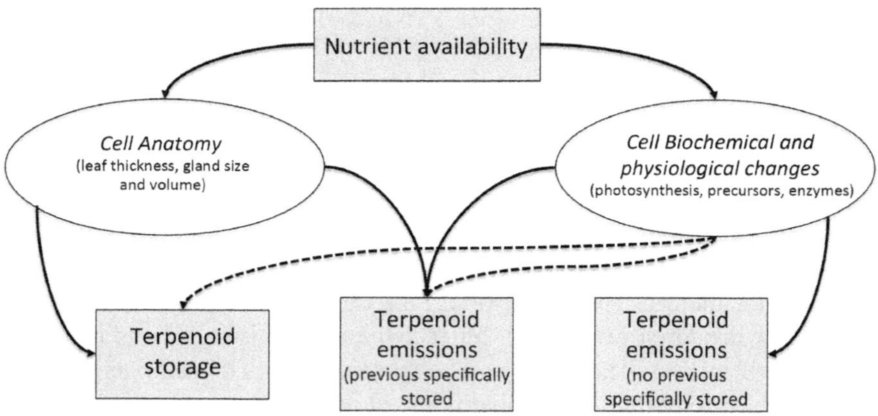

Figure 5.2. *Effect of nutrients on terpenoid storage and emission. Continuous and discontinuous arrows indicate, respectively, the direct and indirect relationship*

This synthesis of secondary compounds has an important cost for the plant and necessitates a trade-off between the allocation of resources for defense and for other functions like growth and reproduction. Usually, the energy cost of defense is presented as a limiting performance of the organism in terms of growth and reproduction. This cost of synthesis and storage of secondary compounds is a complex problem and many theories have been proposed: the carbon-nutrient balance hypothesis, the resource availability hypothesis and the growth-differentiation balance hypothesis (GDBH). The latter, considered as the most integrative, postulates that, when resources increase, growth and biomass will increase constantly while the other processes, like secondary metabolism, will present a bell curve. Secondary metabolism is, therefore, maximal when resources are average: growth is more limited when photosynthesis and the surplus of photosynthetic production are, therefore, allocated preferentially to the differentiation including synthesis of secondary compounds (Figure 5.3(a)). This theory has been partially confirmed several times by using nutrients or water as a resource and by using different types of secondary compounds as a response. For terpenoids' response to hydric resource, Genard-Zielinski *et al.* [GEN 14] have observed an increase in production of secondary compounds with a less-intense stress but with a more limited photosynthesis (Figure 5.3(b)). This type of result shows the complexity of the relationship between primary and secondary metabolisms as a function of species or of resources studied.

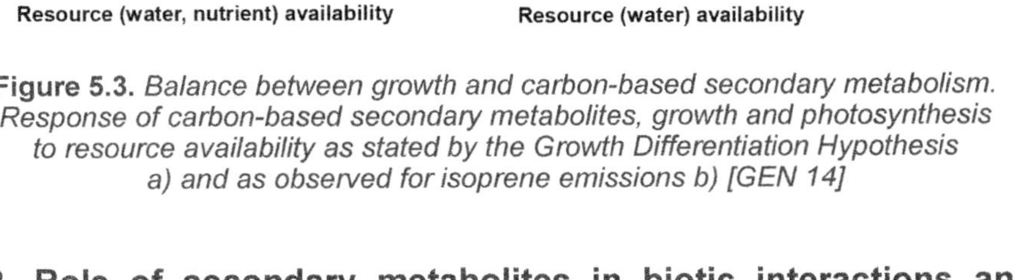

Figure 5.3. *Balance between growth and carbon-based secondary metabolism. Response of carbon-based secondary metabolites, growth and photosynthesis to resource availability as stated by the Growth Differentiation Hypothesis a) and as observed for isoprene emissions b) [GEN 14]*

5.2. Role of secondary metabolites in biotic interactions and community structure

Secondary compounds (also called allelochemicals) produced by organisms are often present in all structures (e.g. leaves, flowers, fruits, roots and stems of a plant). Following their production, they are disseminated into the environment via different release pathways.

In terrestrial environments, the "lightest" lipophilic compounds are mainly released into the atmosphere but also, to a lesser measure, into the soil via volatilization. Hydrophilic compounds, such as phenolic compounds, will be more easily released through foliar or litter leachates. On the other hand, the root exudation and the decomposition of organic plant detritus are also important release pathways (Figure 5.4). Once in the soil, water and/or the atmosphere, the fate and toxicity of allelochemicals are affected by many factors (e.g. nature of the soil).

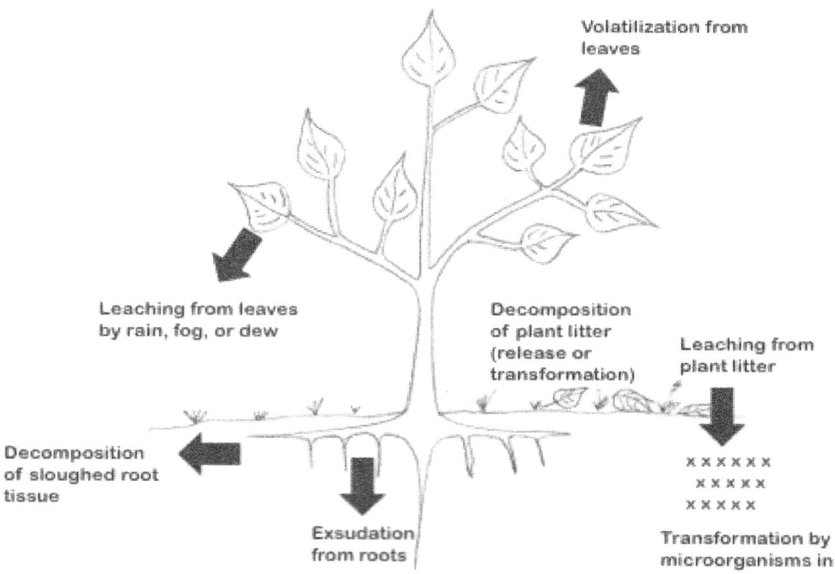

Figure 5.4. *Environmental routes of entry of allelochemicals into environment (from [LAM 08])*

In marine environments, the hydrophilic compounds produced by the sessile or mobile organisms are the easiest to release into the atmosphere, for example, osmolyte dimethylsulfoniopropionate (DMSP) known in the pelagic interactions. Meanwhile, liposoluble components like furanodiene terpenoids are found in water even if their solubility is low. Finally, decomposition will also participate in the release of compounds into this environment [MOL 14].

The biological activity of secondary compounds released into the environment is exerted not only on animals and pathogens in direct contact with the producer but also on other organisms and/or microorganisms which develop nearby. Such interactions are called allelopathic. Allelopathy includes all direct or indirect effects, positive or negative, of a plant (microorganisms included) on another plant by biochemical compounds released into the environment. Allelopathic interactions influence not only natural or agricultural terrestrial ecosystems but also marine or freshwater environments. In these environments, allelopathy is not limited to plant–plant interactions but extends to sessile animals and unicellular organisms. This process could influence the biodiversity of ecosystems, community structures and plant successions. For example, it has been shown by many studies that the Mediterranean secondary succession is influenced by chemical interactions of one key species: *Pinus halepensis* or Aleppo pine. In the northern part of the Mediterranean basin, the abandonment of agricultural lands, observed since the end of the 19th Century, has led to a progression of forestland cover by this species. This expansionist species is rich in secondary metabolites and presents an important allelochemical diversity (e.g. phenolic compounds, fatty acids, mono- and sesquiterpenes), which influence plant communities. Aleppo pine allelochemicals present marked allelopathic potentialities, which will be expressed all along the succession. During abandoned agricultural land colonization, pine, via its allelochemicals, shapes plant biodiversity by eliminating or favoring some species typical of abandoned land (limitation of their germination and growth). Species sensitive to allelochemicals are also absent from last stages of succession (old pine forests). Conversely, non-sensitive species to Aleppo pine allelochemicals will persist in these pine forests. Therefore, there is a link between the sensitivity to allelochemicals and the abundance of species during succession [FER 13]. Later, in succession, when the pines form monospecific forests, allelopathy will express on other species, in particular on the Aleppo pine itself, explaining in part the weak regeneration of this species and the appearance of hardwoods in these formations. This autotoxicity of the dominant species of the forests has been observed elsewhere in many forest ecosystems.

These allelopathic interactions are more difficult to demonstrate than direct plant–animal interactions due to the diffusion of molecules in air, water or soil, and also transformations that these compounds may undergo, notably in contact with organic fractions and minerals of the superficial soil horizons. Thus, numerous interference mechanisms such as the influence of soil and microorganisms or even competition for resources can operate in parallel to the mechanisms of allelopathy. To be effective, these compounds must be i) bioavailable and ii) absorbed by the target plant in sufficient quantities to inhibit germination or any other stage of plant development. Toxic activity can also exert itself while favoring or inhibiting microbial populations notably those involved in the regulation of soil fertility. The contribution of allelochemicals, like phenolic compounds, can lead to changes in soil characteristics like pH, conductivity or potassium soil content. Phenolic acids are also considered as compounds able to greatly influence the nutrient cycles in a terrestrial environment. Complexation with nutrients can change their availability and their turnover in the soil. The observed allelopathic effects can, in some cases, be due to not only the phototoxicity of the compounds but also their action on nutrient availability. One well-described case is that tannins, through complexation with litter nitrogen, block this nutrient under an organic form hardly accessible to non-adapted (micro)organisms. It has also been shown in the laboratory that some phenolic compounds and terpenoids increase immobilization of NH_4^+.

In aquatic environments, the allelopathic processes have been again demonstrated in freshwater and in marine environments. These processes were demonstrated in macrophytes, numerous phytoplanktonic species and also the macrophytes–phytoplankton interactions [HIL 08]. In fact, the inhibition of phytoplankton by allelochemicals of submerged macrophytes is one of the major processes which help the water necessary to maintain shallow lakes. If these allelopathy phenomena were first shown for fixed plants, aquatic environments also contain many sessile species developing chemical mediators involved in their defense and competition considered as allelopathic process. Thus, corral is known for the production of toxic compounds, which participate in the defense of their habitat. For example, *Tubastrea faulkeneri* synthesize indolic alkaloids and other sterols and fatty acids which limit the implantation of concurrent larvae [KOH 00].

Allelochemicals, effective in the natural environment, are also studied for their use in agriculture such as herbicide for the control of weeds or in marine environments to combat the proliferation of toxic algae.

5.3. Secondary metabolites and ecosystem functioning: plant soil relation – brown food chain

Biogeochemical cycles, notably those of carbon and nitrogen, correspond to the circulation of chemical elements from one reservoir to another, that is, biosphere and atmosphere, and to their recycling, altering organic and inorganic forms. These cycles involve organisms, which are, depending on the case, consumers, fixers, transformers and vectors of these elements. Biogeochemical cycles, including organic matter recycling, govern the bioavailability of nutrients, community structure, productivity, energy transfer and the overall ecosystem functioning. In all ecosystems, organic matter is either autochtonous, produced in the ecosystem (i.e. biomass of organisms), or allochtonous, imported from other ecosystems. The relative amount of these two types of organic matter sources depends on the ecosystem and the disturbances they are subjected to. These two main sources of organic matter are equally affected by global changes, including in the least accessible ecosystems (e.g. polar regions for lands or abyssal zones for the sea). This is, for example, an increase in plant production with global warming or an increase in CO_2 concentration (forest cover, phytoplanktonic production) or loss of organic matter in ecosystems (e.g. deforestation, soil erosion).

In terrestrial and freshwater ecosystems, leaf litter represents a large amount of allochtonous matter and constitutes the main energy and organic matter resources for numerous organisms in interaction [HÄT 05]. This organic matter will undergo transformations through the process of decomposition. Litter decomposition, a key process for ecosystem functioning, determines the recycling of all the nutrients from organic to mineral forms available for plant growth. Decomposition is mainly a biological process, involving many groups of organisms, prokaryotes (bacteria) or eukaryotes (fungi and invertebrates). These organisms play an essential role in the transformation of organic matter and its incorporation in trophic webs. The rate of decomposition of organic matter depends on abiotic factors (e.g. temperature, humidity) and also biotic factors such as the type of decomposers and the biochemical composition of the organic matter. Two types of elements enter into this composition: i) the primary metabolites, which integrate the intracellular and storage material (e.g. proteins, glucides), easily degradable and structural compounds (e.g. cellulose, holocellulose, and lignin), more refractory to decomposition and ii) the secondary metabolites, which can be difficult to decompose and also inhibit growth and activity of decomposers.

Plant detritus can, therefore, contain labile compounds (e.g. sugars), refractory compounds to decomposition (e.g. lignin) and decomposers' inhibitor compounds (e.g. phenolic compounds). The relative proportion of these compounds in the litter will influence its rate of decomposition because decomposer organisms will have different abilities to degrade each type of compound. Also, microbial decomposers can generate new compounds from their transformation activity (e.g. tannin–protein complexation).

Concerning the dynamic of these different compounds during the course of the decomposition, three main phases have been identified. During the first phase, the leaching of soluble compounds is the dominant process; therefore, there is a rapid decrease in the litter mass due to a rapid loss of metabolizable and easily leachable compounds. The second phase consists of a combination of fragmentation by soil organisms, a chemical alteration by microorganisms and leaching of compounds which have been transformed and become soluble. This second phase is more controlled by the quality of the litter. The final phase is much slower and involved a chemical modification of the organic matter, which is mixed with the mineral soil and the leaching of litter degradation products from the upper layers; it is often governed by the decomposition of lignin.

Meanwhile, even if the influence of primary metabolites or structural compounds on the decomposition process is now well known, the study of importance of secondary metabolites was until now neglected because of the difficulty of analyzing them, when we know that the palatability of living tissue is directly linked to these secondary metabolites. We know that the biochemistry of green leaves and that of litter is highly correlated [ORM 06] with the physiological and structural characteristics of green leaves, which persist after senescence, and which will, therefore, be directly linked to the process of decomposition, a phenomenon commonly called "the afterlife effect". For example, it has been demonstrated that the concentration of total phenolic compounds in green leaves was positively correlated with the concentration of these compounds in the litter and that these compounds inhibited the colonization of the litter by fungi (Figure 5.5).

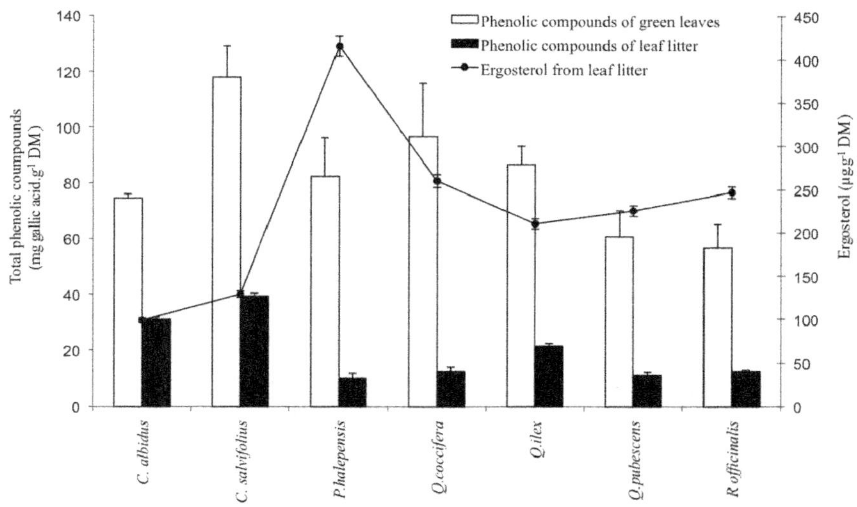

Figure 5.5. *Total phenolic compound concentrations (in mg of gallic acid.g^{-1} DM) of green leaves, total phenolic compound (in mg of gallic acid.g^{-1} DM) and ergosterol concentrations (µg.g^{-1}DM) of leaf litter of the seven species studied. Bars denote 95% confidence limit (N = 3) [ORM 06]*

Globally, these phenolic compounds are poorly degradable and can limit the growth and activity of decomposer organisms, from macro-arthropods to microorganisms. Thus, they will be important regulators of the carbon and nitrogen cycle by slowing down the rate of litter decomposition [ORM 06].

Certain phenolic compounds, such as caffeic acid and P-coumaric acid, can determine the decomposition of Aleppo pine needles in secondary successions, with a slowdown in the colonization needles by decomposers in young pines, in comparison to older pines (Figure 5.6) due to longer residence time for phenolic compounds in pine needle litter.

Tannins can also affect the nitrogen and carbon cycles; they can complex proteins or metallic ions and can present toxicity for microorganisms and inhibit enzymatic activities. In the majority of studies, mineralization of the nitrogen is decreased by the addition of tannins, while the effects on carbon mineralization remain more variable. By their characteristic link to proteins, tannins inhibit the microbial extracellular enzymes and can also limit the available substrates for microbial growth. The difference in tannin content of the leaves seems to act on the aquatic macroinvertebrate communities using poplar leaves falling in adjacent watercourses as resources, and on the mineralization of the nitrogen or the decomposition of these litters in aquatic environments.

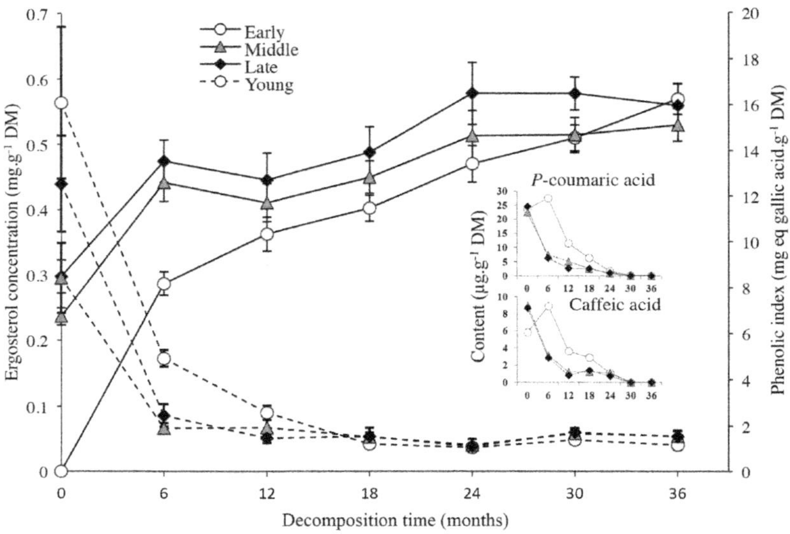

Figure 5.6. *Dynamics of ergosterol litter content (solid lines) and phenolic index (dashed lines) over the course of decomposition (means and SD). Inset: change in two specific phenolic compound litter contents (caffeic and p-coumaric acid) during decomposition in the three successional stage plots [CHO 14]*

Lignin, an organic substance synthesized by plants, the most abundant after cellulose, is a polyphenol whose role in the decomposition process is largely known. It has been used much more in ratio form (lignin/N) and appears as a fundamental characteristic for the prediction of the rate of decomposition of different litters. Lignin is resistant to decomposition by reason of its aromatic rings and its high degree of polymerization; only ligninolytic fungi are able to synthesize extracellular enzymes which decompose these structures and transform them into forms assimilable by other organisms. However, and beyond these characteristics, lignin does not really have any biological activity, and it is often considered as a primary metabolite.

Finally, the impact of terpenes on soil microorganisms, in particular the monoterpenes, is complex, because it can inhibit the activity and growth of certain microbial groups, and stimulate others. However, studies have shown that monoterpenes inhibited the mineralization of N and the net nitrification in the soil. Causes of these inhibitions are still not fully understood, but may be due to a direct action of monoterpenes on an enzyme involved in the ammonium oxidation pathway, or to an inhibition of the growth in *Nitrosomonas europaea*, a bacterium that oxidizes ammonium to nitrite (nitrification). These authors also noted that other terpenes, such as beta-pinene, seem to have stimulatory effects on growth *of Nitrosomonas europaea*.

White [WHI 91] proposes a response model of inorganic nitrogen in soil with the addition of monoterpenes. At very low concentrations, these compounds induce a slight increase in mineralization. At a higher concentration, the authors observed an increase in the proportion of ammonium in comparison with nitrates, translating to an inhibition of nitrification. With a still higher concentration, they noted immobilization of ammonium, representing the only source of mineral nitrogen still present in the soil (Figure 5.7).

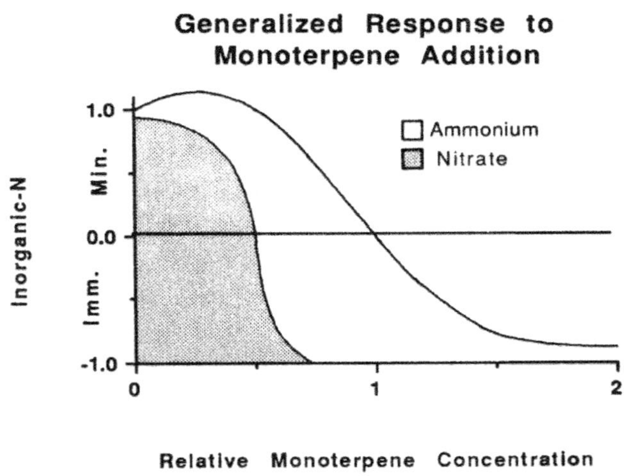

Figure 5.7. *Figure, from [WHI 91], modeling the generalized response of relative concentration of soil inorganic nitrogen to the addition of monoterpenes. Mineralization (Min.) occurs when the quantity of inorganic nitrogen increases during incubation. Immobilization (Imm.) occurs when the total quantity of soil inorganic nitrogen falls below initial concentrations. The contributions of monoterpenes are in relative values and not concentrations*

In turn, secondary metabolites serve as sources of carbon for decomposer organisms. Thus, microorganisms can play an important role in the detoxification of litter by the degradation of these secondary metabolites. Tannins of low molecular weight can act as carbon substrates for microorganisms when they are adapted to this type of compound. In fact, microbial communities which have had a limited exposure to tannins during their development seem to be more sensitive to these compounds.

The chemical quality of litters can also greatly influence communities of soil arthropods. A study of litter decomposition *in situ* in a Mediterranean forest highlighted the structuring of communities of springtails as a function of litter type: *Cotinus coggygria* had the lowest abundance and diversity, and *Pinus halepensis* had the highest diversity with certain species of springtails preferentially associated with this litter. *In vitro* ecotoxicological tests on *Folsomia candida* (ubiquitous species of springtail) highlighted a toxic effect of aqueous extracts of *C. coggygria* litter, and, on the opposite, a positive effect of aqueous extracts of *P. halepensis* litter on the survival and reproduction of this species. Finally, olfactometry tests showed an attraction of *Folsomia candida* for the litter of *P. halepensis*. These results obtained in the laboratory contribute to understanding the key role of secondary metabolites in the structuring of soil anthropod communities [SAN 14]. Meanwhile, simulations of the increase in summer drought, foreseen in the context of climate change in Mediterranean environments, show modifications of the relation between litter chemical diversity and the structure of arthropods communities (Figure 5.8).

Finally, litter chemical diversity (across the diversity of plant species) can be directly linked to the diversity of decomposers and, therefore, to the decomposition efficiency. The hypothesis is that the interactions between the different litters come from, in part, the effects of resources heterogeneity on fungal, bacterial and arthopod activity. It has been shown that, in certain cases, an increase in chemical diversity of the litter increases the respiration and N-mineralization or litter decomposition. Meanwhile, the quality of some plant species present in litter mixtures overrides the chemical difference between these litters in the control of the process.

Figure 5.8. *Experimental forest (O_3HP: Oak Observatory at "l'Observatoire de Haute Provence") with a rain exclusion system allowing climatic change simulation in a Mediterranean environment (increase in summer drought) (photo: Thierry Gauquelin)*

5.4. Integration of biotic and abiotic dynamics: benthic marine microhabitats

The integration of biotic and abiotic dynamics also needs to take into account interactions between organisms and chemical properties of the environment, which are structured, not only by the synthesis of metabolites, but also by local disturbances of the physicochemical equilibria. The study of organic matter degradation processes, in the form of particulate or macrodebris, in benthic marine environments, combined with that of community diversity and dynamic, helps illustrate this integration.

Contrary to physical drivers of the environment, such as temperature, light, pressure or salinity, which are exerted homogeneously over large scales, physicochemical factors of the marine environment are modulated locally by the activity of biological communities. This is the case, for example, of oxygen, whose concentration can vary from its level of atmospheric saturation to zero, in several millimeters or less in marine sediments rich in organic matter (Figure 5.9).

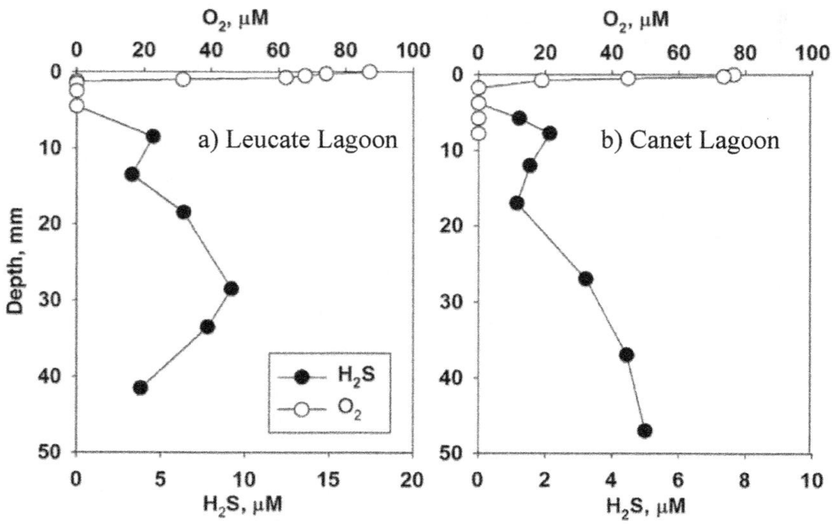

Figure 5.9. *Profile of sulfide and oxygen in the sediment of two coastal lagoons, illustrating the shallow depth of oxygen penetration in sediments rich in organic matter and the sulfur enrichment of upper sediment layers (from [YÜC 13]).*

Benthic microbial communities govern the position of this interface between the oxic environment (upper sediment layer and the water column) and the anoxic sediment layers. The spatial variability of oxygen profiles, from coastal zones to great oceanic depths, reflects the intensity and the degradability of detritic organic contributions. These vertical oxygen profiles, at the water–sediment interface, are used by biogeochemists to establish the rate of remineralization of sedimented organic matter, which can be produced in the water column or transported by rivers. Besides the storage of carbon, these processes also play a major role in the functioning of ecosystems. In fact, the major resource of benthic marine communities is constituted by these detritic contributions, from the primary production of other ecosystems to which the sediment compartment is associated. The primary production is planktonic for most oceanic systems, but can be dominated by terrestrial plants, such as mangroves, or chemosynthetic plants on continental margins rich in hydrocarbons or in the hydrothermal environments.

A temporal dynamic is most often associated with this spatial variability of detritic contributions, either seasonal or consecutive of the modification of flux of organic matter by human activities (e.g. linked to the aquaculture in the sea). The biogeochemical approach of marine sediments considers these systems as a succession of stationary states but rarely takes into account the nonlinear responses of microbial communities to the variability of their environment. The stability of this system meanwhile depends on the capacity of communities to maintain or reestablish the sediment surface oxic conditions required for maintaining the diversity of the benthic fauna. On the other hand, anaerobic communities predominate in the degradation of organic matter by using as electron acceptors nitrates and especially sulfate from seawater, and by producing reduced inorganic compounds like sulfide or ammonium. These reduced inorganic compounds are not only toxic for aerobic organisms but can react in an abiotic manner with oxygen, amplifying its depletion in the environment and sometimes in the surface water. It, therefore, appears indispensable to understand the dynamic of these benthic systems, by integrating the dynamic of biotic interactions and their feedback on the physicochemical conditions of the habitat.

The influence of organisms on the abiotic environment and the reestablishment of stationary conditions pass through non-stationary states for which the abiotic characteristics of the habitat cannot be dissociated from the dynamics of the biological processes. For example, the appearance of chemolithotrophic microorganisms regulates the H_2S content produced by the degradation of vegetable debris in the marine environment [YÜC 13], establishing feedback loops which favor the colonization of these substrates by invertebrates while reducing habitat toxicity [LAU 09]. Chemolithoautotrophic microorganisms constitute a source of new organic matter, which is rarely taken into account, except in environments where this mode of primary production predominates such as hydrothermal vents and sediments rich in methane [ORC 11]. The role of hydrodynamics or sedimentary instability for the establishment of such systems in common marine environments just begins to be taken into account (e.g. for fluidic mud off the coast of the Amazon or the sandy sediments subject to the flow of oxygen-saturated water under the influence of tides and currents) or, at a smaller scale, on plant debris degrading in mangroves (Figure 5.10), even if the responses of the microorganisms to these local instabilities have not yet been fully identified.

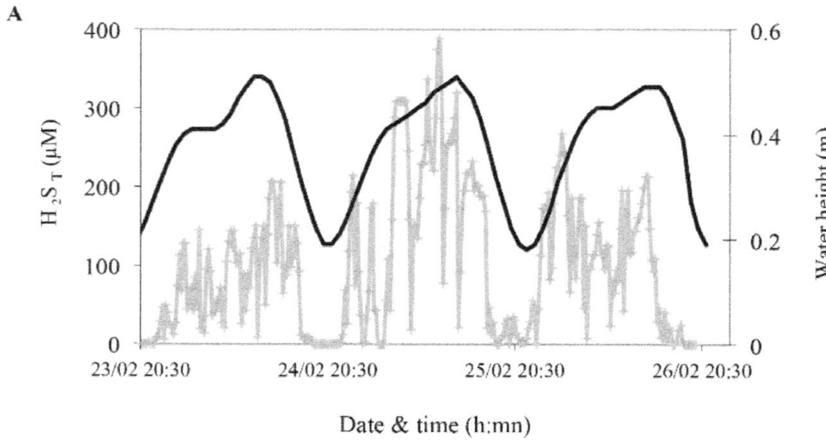

Figure 5.10. *Variability of sulfide concentrations (gray curve) to the surface of wood debris immersed in mangrove water under the influence of tidal currents (black curve) (from [LAU 09])*

The study of the dynamics of biogeochemical profiles governed by microorganisms at benthic interfaces (e.g. wood–water, sediment–water, microbial mat–water; Figure 5.11) constitutes one of the keys for understanding the relationships between benthic diversity, the degradation of organic matter and responses to disturbances at different scales, either climate-driven or linked to direct impacts on the seabed.

It is, however, not sufficient, and the activity of the diverse benthic communities as a whole, including all components of the fauna and flora, and hence the physical or chemical heterogeneity of microhabitats, also has to be considered. In this context, benthic invertebrates, the engineers of the ecosystem, play a major role. Certain species of endogenous macrofauna can attain important densities, favor the exchange between oxidant and reducer compounds through the walls of their burrows and regulate the microbial metabolic functions in these sediment microniches, which differ from the classic "redox horizons" (Figure 5.11). Molecular approaches in environmental genomics now complement the toolbox for characterizing sedimentary micro-profiles, providing access to the diversity and temporal variability of microbial communities associated with these habitats [BER 09]. These microscale phenomena have effects on large-scale ecosystem collection. The bioirrigation of burrows not only regulates the oxygen content of sediment and favors the aerobic degradation of organic matter, but also increases the flux of remineralized nitrogen to the pelagic ecosystem. The fluctuations of populaions of different species of annelids generate modifications of the surface phytoplanktonic productivity in environments characterized by a strong benthic–pelagic coupling [KRI 14]. Plants, notably seagrasses or mangroves, also exert control on the oxygenation of sediment across their rhizosphere, modifying the transformations of organic matter composed of plants debris.

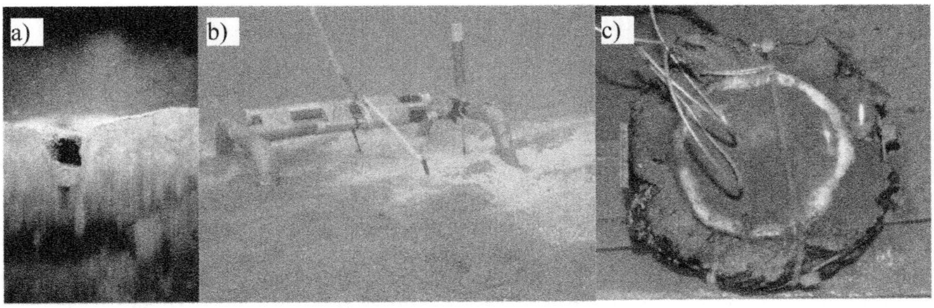

Figure 5.11. *Chemical heterogeneity of microhabitats: a) burrow promoting the diffusion of oxygen in prodelta of sediments of the Rhone (image SPI, LECOB); b) sulfo-oxidizing microbial mats at the surface of hydrothermal sediments equipped with sensors (image LECOB-Rutgers University); c) measure of sulfide at the surface of wood immersed in seawater (image from LECOB)*

For marine ecosystems subjected to multiple pressures, such as coastal ones, the necessity to integrate all the reactive components of the system, chemical and biological, is particularly important to predict the response to hypoxia phenomena produced by eutrophication of the environment. Understanding the consequences of enhanced hypoxic episodes in coastal regions requires considering, not only the responses of different species at different thresholds of exposure (lethal, sublethal) at different stages of their lifecycle, but also their influence on oxygen content and sulfide production in its most toxic form, H_2S, in the habitat [VAQ 10]. Synergistic effects between physical disturbances (increase in temperature, sedimentation) and sulfide production can increase the mortality of engineer species. For example, the impacts of warming waters on seagrass induce a biogeochemical cascade leading to a spatial increase in mortality zones within the seagrass bed resulting in an increase in the sulfide concentration in the sediment.

5.5. Conclusions

These examples illustrate the fact that the reactive biotic and abiotic compounds forming the ecogeochemical network cannot be dissociated. The understanding of control processes exerted on these dynamics by microorganisms and metazoans is still very patchy, and differs greatly according to the ecosystems studied, terrestrial or marine, and the disciplinary domains which have helped the emergence of these approaches: biology, chemistry or geochemistry. Interdisciplinary approaches, notably in environmental genomics, offer opportunities for integrating the understanding of different types of ecosystems and studying of fundamental features. The current tools also help to understand dynamic couplings, which establish between components which govern the responses. The integration effort must also involve the interaction networks considered in a non-stationary context, and on multiple spatial scales which combine the scales of the organism to that of the ecosystem and of the microhabitat to the landscape. Ecogeochemistry proposes to analyze, by integrative approaches, the complexities of ecological systems and the mechanisms by which biotic and abiotic compounds of the ecosystem interact. They complete the classical approaches of functional ecology by supporting on the same plan the organisms and components of their abiotic environment, particularly the chemical components in interaction with these organisms.

5.6. Bibliography

[BER 09] BERTICS V.J., ZIEBIS W., "Biodiversity of benthic microbial communities in bioturbated coastal sediments is controlled by geochemical microniches", *ISME Journal*, vol. 3, no. 11, pp. 1269–1285, 2009.

[CHO 14] CHOMEL M., FERNANDEZ C., BOUSQUET-MÉLOU A. *et al.*, "Secondary metabolites of *Pinus halepensis* alter decomposer organisms and litter decomposition during afforestation of abandoned agricultural zones", *Journal of Ecology*, vol. 102, pp. 411–424, 2014.

[FER 13] FERNANDEZ C., SANTONJA M., GROS R. *et al.*, "Allelochemicals of *Pinus halepensis* as drivers of biodiversity in Mediterranean open mosaic habitats during the colonization stage of secondary succession", *Journal of Chemical Ecology*, vol. 39, no. 2, pp. 298–311, 2013.

[GEN 14] GENARD-ZIELINSKI A.-C., ORMEÑO E., BOISSARD C. *et al.*, "Isoprene emissions from downy oak under water limitation during an entire growing season: what cost for growth?", *PLos One*, vol. 9, no. 11, p. e112418, 2014.

[HÄT 05] HÄTTENSCHWILER S., TIUNOV A., SCHEU S., "Biodiversity and litter decomposition in terrestrial ecosystems", *Annual Review of Ecology, Evolution, and Systematics*, vol. 36, pp. 191–218, 2005.

[HIL 08] HILT S., GROSS E.M., "Can allelopathically active submerged macrophytes stabilise clear-water states in shallow lakes?", *Basic and Applied Ecology*, vol. 9, no. 4, pp. 422–432, 2008.

[HOP 03] HOPKINS W.G., *Physiologie végétale*, De Boeck, 2003.

[KOH 00] KOH E.G.L., SWEATMAN H., "Chemical warfare among scleractinians: bioactive natural products from Tubastraea faulkneri Wells kill larvae of potential competitors", *Journal of Experimental Marine Biology and Ecology*, vol. 251, no. 2, pp. 141–160, 2000.

[KRI 14] KRISTENSEN E., DELEFOSSE M., QUINTANA C.O. *et al.*, "Influence of benthic macrofauna community shifts on ecosystem functioning in shallow estuaries", *Frontiers in Marine Science*, vol. 1, 2014.

[LAM 08] LAMBERS H., CHAPIN III F.S., PONS T.L., "Ecological biochemistry: allelopathy and defense against herbivores", in LAMBERS H., CHAPIN F.S., PONS T.L. (eds), *Plant Physiological Ecology*, Springer, New York, 2008.

[LAU 09] LAURENT M.C.Z., GROS O., BRULPORT J.-P. *et al.*, "Sunken wood habitat for thiotrophic symbiosis in mangrove swamps", *Marine Environmental Research*, vol. 67, pp. 83–88, 2009.

[MOL 14] MOLLO E., FONTANA A., ROUSSIS V. *et al.*, "Sensing marine biomolecules: smell, taste, and the evolutionary transition from aquatic to terrestrial life", *Frontiers in Chemistry*, vol. 2, pp. 92, 2014.

[ORC 11] ORCUTT B.N., SYLVAN J.B., KNAB N.J. *et al.*, "Microbial ecology of the dark ocean above, at, and below the seafloor", *Microbiology and Molecular Biology Reviews*, vol. 75, pp. 361–422, 2011.

[ORM 06] ORMEÑO E., BALDY V., BALLINI C. *et al.*, "Effects of environmental factors and leaf chemistry on leaf litter colonization by fungi in a Mediterranean shrubland", *Pedobiologia*, vol. 50, pp. 1–10, 2006.

[ORM 12] ORMEÑO E., FERNANDEZ C., "Effect of soil nutrient on production and diversity of volatile terpenoids from plants", *Current Bioactive Compounds*, vol. 8, pp. 71–79, 2012.

[PIC 11] PICHERSKY E., LEWINSOHN E., "Convergent evolution in plant specialized metabolism", *Annual Review of Plant Biology*, vol. 62, pp. 549–566, 2011.

[SAN 14] SANTONJA M., Impact du changement climatique sur le processus de décomposition des litières en région méditerranéenne française, Doctoral Thesis, Aix-Marseille University, 2014.

[VAQ 10] VAQUER-SUNYER R., DUARTE C.M., "Sulfide exposure accelerates hypoxia-driven mortality", *Limnology Oceanography*, vol. 55, no. 3, pp. 1075–1082, 2010.

[WHI 91] WHITE C.S., "The role of monoterpenes in soil nitrogen cycling processes in ponderosa pine – results from laboratory bioassays and field studies", *Biogeochemistry*, vol. 12, pp. 43–68, 1991.

[WIL 89] WILLIAMS D.H., STONE M.J., HAUCK P.R. *et al.*, "Why are secondary metabolites (natural products) biosynthesized?", *Journal of Natural Products*, vol. 52, no. 6, pp. 1189–1208, 1989.

[YÜC 13] YÜCEL M., "Down the thermodynamic ladder: a comparative study of marine redox gradients across diverse sedimentary environments", *Estuarine Coastal Shelf Science*, vol. 131, pp. 83–92, 2013.

6

Omics in Chemical Ecology

Several years ago, biology entered a new era, one of high-throughput technologies, which, by simultaneously allowing for the analysis of a large number of genes, transcripts, proteins and metabolites, revolutionized the scale of analysis and provided a more detailed understanding of complex biological mechanisms. From this technological progress, new approaches have emerged, often captured under the term "omics" (a term which probably comes from the Sanskrit term "OM" denoting completeness), which use these high-throughput techniques to study living molecules at a very large scale and in an integrative manner. In the field of chemical ecology, the involvement of "omics" approaches goes beyond the simple methodological contribution: the term "omics" increasingly establishes itself as a research field in its own right (emergence of the term "ecogenomics"), which is becoming indispensable with the growing democratization of high-throughput sequencing analyses. The "omics" approaches meanwhile only represent one step in response to the issues of chemical ecology and have already given rise to new questions. For example, thanks to metagenomics, it is now possible to understand phenomena directly in their natural environment. In addition, the application of "omics" approaches sheds new light on evolutionary mechanisms. Finally, the post-genomic step will help to elucidate the function of genes, from their genetic regulations to their interactions in complex networks. Research topics using "omics" approaches in chemical ecology are quite diverse and based on various biological models from microorganisms to mammals, plants and insects (Figure 6.1).

Chapter written by Sylvie BAUDINO, Christophe LUCAS and Carole SMADJA.

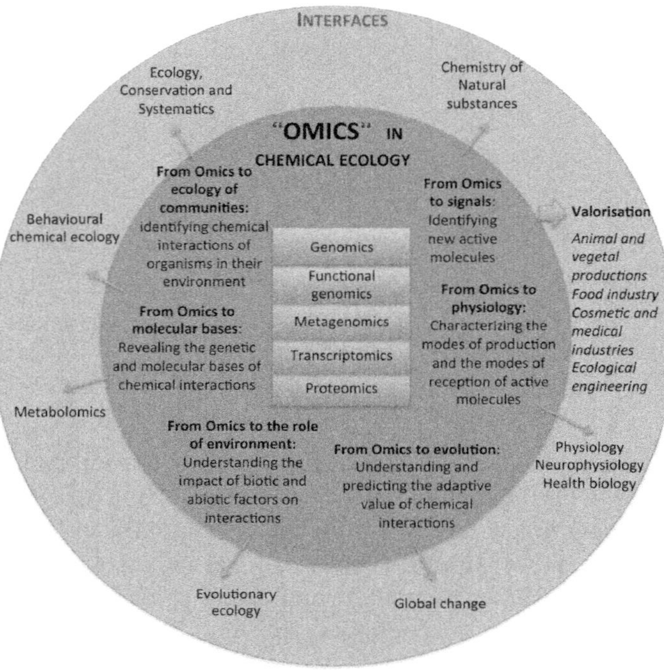

Figure 6.1. *The "omics" revolution in chemical ecology*

The different themes above are often isolated but this is because of the multitude of biological models and the associated scientific questions, such as the separation between "terrestrial" and "marine" chemical ecology. In this context, the chapter will take the form of responses to the questions with a common scientific approach and which are associated with characteristic examples of the different models studied. In this chapter, examples were mainly chosen among macroscopic eukaryotes, Chapter 4 more specifically dealing with the microbiome. In addition, metabolomic approaches, being the subject of Chapter 7, will only be very briefly mentioned.

6.1. Introduction: the different "omic" technologies

The so-called "omic" approaches simultaneously study a large number of genes, transcripts, proteins and metabolites, without *a priori* knowledge of their biological functions (Figure 6.2). Genomics is the large-scale study (entire genome or important fraction of the genome) of the coding or non-coding genetic material of an organism. When we study the genetic material of a sample consisting of numerous organisms coming from a complex environment (gut, ocean, soil, etc.), we speak of metagenomics. This approach is notably used to address the classification of the microorganisms' species which are difficult to identify, or for the functional analysis of the different players in an ecological community. In a similar manner, transcriptomics, proteomics and metabolomics help to identify and analyze all the transcripts, proteins and metabolites constituting the transcriptome, proteome and metabolome of one or more organisms in a given physiological state or environment. Mass data production techniques, such as the double hybrid method, are also used for the study of interactions between proteins (interactomics).

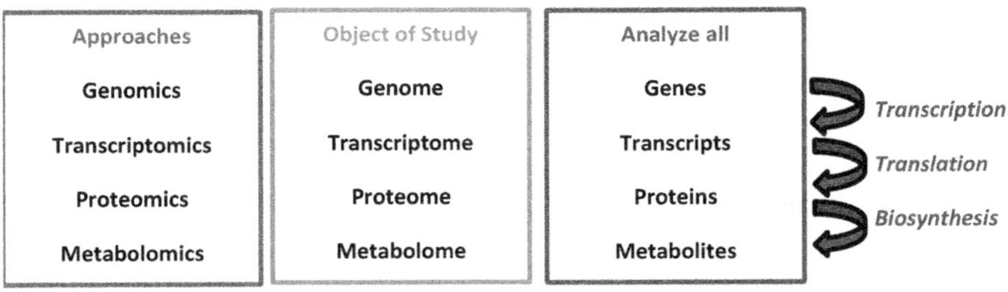

Figure 6.2. *Main "omic" approaches and their links at the scale of a living organism*

These large-scale approaches are possible thanks to the development of high-throughput innovative techniques (DNA chips, massive sequencing (Next-Generation Sequencing or NGS), high-throughput genotyping; see Chapter 8) which, because of their increasingly small associated costs, enable their application to a growing number of projects. Furthermore, some of these techniques (notably NGS sequencing) can be used in non-model organisms, with no detailed information on their genomes. Owing to this, "omic" approaches can be developed in most species, allowing considerable progress in disciplines essentially centered on the diversity of living organisms, as chemical ecology. In the first place, high-throughput techniques allow for the exhaustive identification and cataloging of living molecules (genes, transcripts, other nucleic sequences, proteins, metabolites). They are often used to identify genes of interest involved in a particular function, functional networks and genetic changes underlying the evolution of living organisms. For example, DNA chips or microarray techniques help to measure the differential accumulation of transcripts between a control sample and one or more treated samples. A more recent technique, which relies on novel high-throughput sequencing technologies, is the sequencing of entire transcriptomes (called RNA-seq), which allows for the identification of all transcripts of a given organism or organ and the analysis of their levels of expression.

Genes identified by these high-throughput techniques are then classified in detail by various methodologies to which biologists have access today. For example, a more detailed analysis of the expression of these genes (qPCR), or the effect of their overexpression or their extinction (RNAi, RNA interference techniques) on the phenotype of the studied organisms, can be considered. Detailed localization of the gene expression in the different organs and tissues can be studied by RNA hybridization *in situ*. Finally, the coupling of proteins of interest with a fluorescent protein such as the Green Fluorescent Protein (GFP) facilitates their visualization in the subcellular compartments.

6.2. From "omics" to signals: identifying new active molecules

In chemical ecology, one of the first applications of "omic" techniques concerns the systematic search for chemical signals involved in biotic interactions, which opens the door to the discovery of new active molecules. This is mainly the subject of metabolomics, which is developed in Chapter 7, but it is also possible to use other "omics" such as proteomics or transcriptomics. Analyses without *a priori* knowledge have been applied to numerous organisms. By heterologous expression of transcripts, directly isolated from the environment of bacteria or soil or ascidia, new signals can thus been searched for. The systematic search for volatile compounds emitted by a tomato was performed to find the molecules which enable the *Cuscuta* parasitic plant to find its tomato host [RUN 06]. The study of this type of plant–plant interaction, called allelopathy, is very new and dozens of identical procedures for plant–insect relations can be cited. The targeted analyses on a family of molecules known to play a signal role, like antibiotics, have also been performed in microorganism communities by metagenomics.

Another method to identify significant biological signals is to first study the emission systems. In Drosophila, new transcripts corresponding to signals involved in communication between insects have been researched systematically in the genome. This is the case for transcripts corresponding to elongases, enzymes that build long-chain fatty acids, or desaturases, enzymes that add double carbon–carbon bonds to certain fatty acids [CHE 07]. These genes have also been identified in non-model species like *Formica exsecta* ants by sequencing of the entire transcriptome. Another approach for discovering new transcripts is to target the organ which emits the signals; for example, all moth transcripts involved in the production of pheromonal signals were characterized [VOG 10]. The same procedure was used on the digestive tube or fat of insect pests like pine bark beetles. The study of soil bacterial competitors and the search for new transcripts have also enabled the discovery of unknown signals [GAR 11]. In bees, it is the reception of signals which have been targeted, by comparing proteomes of the antennas of males and workers and the annotation of genes involved in the detection of odors [FEN 11]. In termites, an enzyme, previously identified as playing a role in the digestion of wood, was revealed, after differential analysis among many castes, as the major actor of inter-individual interactions during the emergence of new reproducers [KOR 09]. The species of *Trichoderma* are fungi which interact with plant roots, like maize, and increase the resistance of these plants to bacterial infections. Owing to this, these symbiotic organisms are increasingly used as biological control agents. The interaction between the fungus and the plant root is mediated at least in part by secreted proteins which have been recently identified by proteomic studies [LAM 15] (see Chapter 4).

6.3. From "omics" to the ecology of communities: identifying chemical interactions of organisms in their environment

In complex communities, many chemical signals are emitted or simply present in the environment. The "omics", after having detected a signal during an interaction, allow for the identification of partners of this interaction in a complex chemical environment. This question, which concerns mainly metagenomics of communities of microorganisms, is developed in Chapter 4. Associations of organisms with bacteria are innumerable but difficult to study. In fact, there are generally a large number of interacting partners and many bacteria are impossible to cultivate. Metagenomics is the preferred approach for studying these complex communities. For example, it helps to distinguish symbionts from other partners of interaction among bacteria of an organism, for example, a sponge. Another example is the study of the origin of a polyketide, coming from symbiotic bacteria, in a beetle [PIE 02], by metagenomics and heterologous expression. Similarly, in the lichen of the genus *Peltigera*, recent metagenomic studies, coupled with other analyses, have highlighted polyketide compounds, until now unknown in lichens, and showed that biosynthesis was performed by the photobiont of the symbiosis, an alga of the *Nostoc* genus [KAM 13]. The presence, already known, of this type of compound in animal–bacteria associations, suggests that they could play a role in symbiosis. A metagenomic study on tobacco in a natural environment showed that colonization by bacterial communities associated with roots and leaves involved different species and that the bacterial flora of the soil in which the plant lived played a crucial role in this colonization. Roots of Spermaphyte plants and fungi of the genus *Glomus* frequently associate to form mycorrhizae. The establishment of this symbiosis involves a complex developmental program, which has extensively been studied by the different "omic" technologies. In the tomato, for example, transcriptomic experiments, coupled with target analysis by laser microdissection, helped to target genes induced by colonization from fungi, in particular auxin pathway genes [FIO 09]. Fungi synthesize numerous volatile compounds, the roles of which are poorly understood. Except in certain organisms of interest like yeasts, the mechanisms of their biosynthesis are still rarely studied. Recently, the genome of an emblematic species, the black truffle, has been sequenced. Following this sequencing, candidate genes of biosynthesis pathways, intervening in particular in the production of sulfurous compounds, were proposed. More recently, it has been shown in another truffle species that the biosynthesis of certain compounds such as thiophenes was not due to the fungus itself, but to symbiotic bacteria [SPL 14].

Finally, many recent works on gut bacterial communities in mammals, metagenomic studies, have shown that the equilibrium of these communities was dependent on numerous genetic and environmental factors.

6.4. From "omics" to molecular bases: revealing the genetic and molecular bases of chemical interactions

This is probably the field of chemical ecology in which the "omics" allowed the most significant advances in the recent years. There are numerous articles presenting an "omic" study leading not only to the partial or complete understanding of a biosynthesis pathway of a secondary metabolite, a pheromone, a volatile compound or any other chemical signal, but also to the characterization of the molecular bases of signal reception.

In microorganisms, "quorum sensing" is a communication system among bacteria, which allows them to regulate their growth and that of their competitors, depending on population density. Through a decentralized system, bacteria are capable of standardizing their biological responses in proportion to their mutual interactions. In this framework, "omic" studies led to the identification of proteins intervening in this phenomenon.

In plants, elucidation of biosynthesis pathways of volatile compounds emitted by flowers, in line with their interactions with animals, is a very active domain. Thus, for example, comparative transcriptomic studies among different fragrant or non-fragrant varieties allowed rapid progress. This is the case of the biosynthesis pathways of 3,5-dimethoxytoluene in horticultural roses (Figure 6.3) decrypted through EST (Expressed Sequence Tag, fragments of ARN partially sequenced during certain transcriptomic studies). It has been shown that this biosynthesis pathway appeared in wild Chinese roses probably by duplication and neofunctionalization of a gene coding for an *O*-methyl transferase [SCA 08].

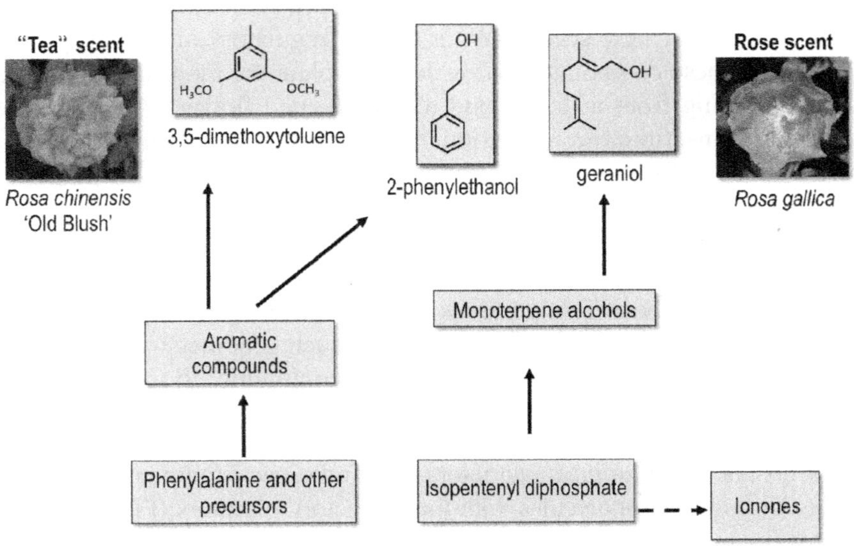

Figure 6.3. *Main biosynthesis pathways of volatile compounds in roses (by S. Baudino and P. Hugueney)*

Other biosynthetic enzymes have also been discovered by proteomic analysis, by EST or by subtractive hybridization in numerous plants, such as the petunia, carnation or snapdragon. Some orchids such as *Ophrys* present very sophisticated pollination systems with a specialization of insects by species. It has been shown that the scent of these flowers, which mimic the sexual pheromone produced by the female insect, is the key attraction factor of specific pollinators (see Chapter 1). This "pseudo-pheromone" is a mixture of hydrocarbons, with a ratio of specific alkanes and alkenes for each species, the biosynthesis pathways of which are not yet elucidated in the orchids. The transcriptomic and proteomic tools helping to identify the intervening genes in these biosynthesis pathways are just beginning to be studied in these emblematic species. In petunia, a comparative analysis by DNA chip helped to identify the *ODORANT*1 gene coding for a transcription factor regulating the biosynthesis of phenylpropanoids. This MYB-type factor was the first regulator of the biosynthesis pathways of volatile compounds identified in plants [VER 05]. Since this time, many others have been described and transcription factors, also regulating the production of non-volatile compounds such as nicotine, have been identified. It is certain that not all pathways are known, still many remain enigmatic and, above all, numerous questions remain as to their functioning. Facing a selection pressure due to pollinators or pests, what are the genetic mechanisms leading to the emission of new scents? What are the regulators of the biosynthesis pathways of these compounds? How do we explain the modification of scent during blooming from a flower bud to a pollinated flower? Although there are many transcriptomic, proteomic and even metabolomic studies on flowers, except for the one of petunia, not many genome sequences of scented flowers are available. These sequences could be invaluable for solving the previous questions. Numerous species are candidates for sequencing, among them are *Mimulus*, tobacco and rose. Recently, many consortiums have been formed to work on sequencing strategies of several fragrant flowers. It is, therefore, possible that such genomes will be available in the near future and will help the scientific community to respond to more general chemical ecology problems.

Insects and other animals can interact with their environment by detecting thousands of odorant molecules with their olfactory receptors (Figure 6.4) (see Chapter 3). Olfaction is a complex procedure that involves the fixation of these odorant molecules on particular proteins present in the olfactory organs: the olfactory receptors (OR). In animals, numerous studies also use "omics" to identify OR and other genes involved in odor reception (OBP or Odorant Binding Proteins, degradation enzymes, etc.).

These studies also seek to highlight the transcription factors regulating signal reception.

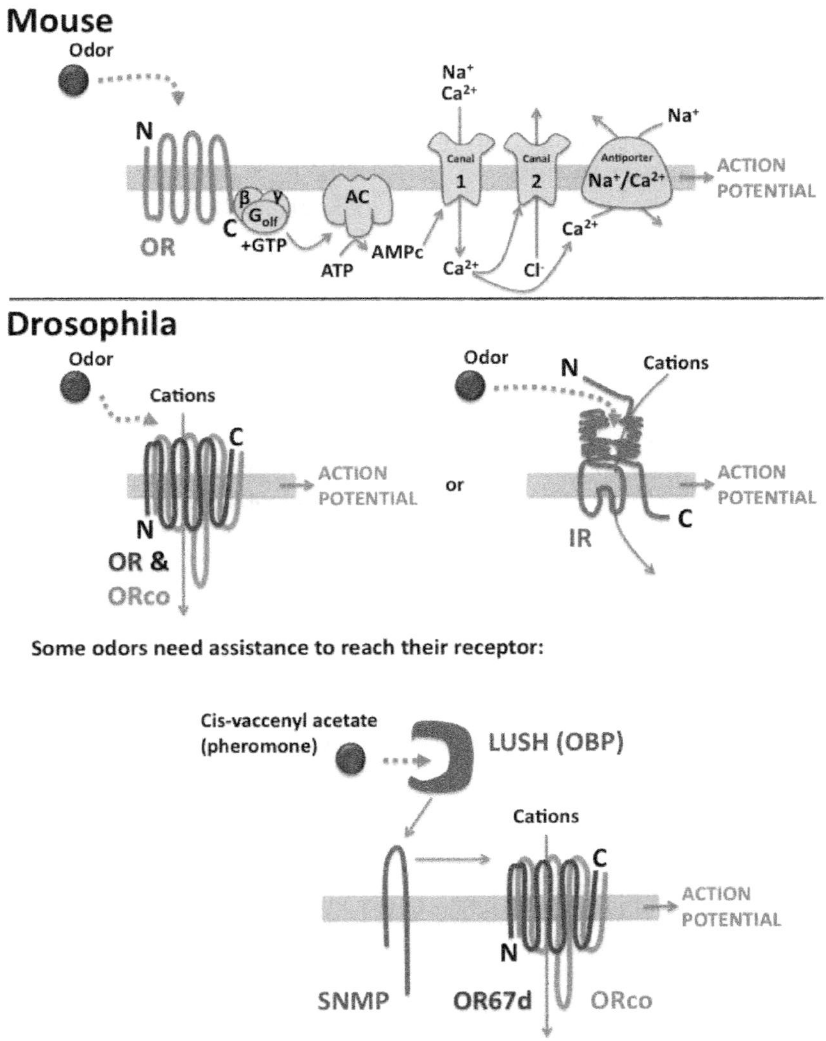

Figure 6.4. *Molecular detection of olfactory signals in mice and drosophila (by Y. Grosjean)*

Since their initial discovery in Drosophila at the beginning of the 21st Century, odorant receptors of insect have been the subject of numerous studies, with the help of various "omic" techniques. The possibility of finding antagonists of these OR, which could be used as methods to control insect pests, has certainly been a stimulating factor of this research. We can cite the identification of odorant receptors and the associated cellular response elements (ionic channels) in the antennae of a butterfly of the *Spodoptera* genus by RNA-seq [LEG 11]. The same technique helped the exhaustive cataloging of odorant and gustatory receptors in butterflies of the *Heliconius* genus, while identifying the odorant receptors involved in specific recognition functions in females and males [BRI 13]. In the *Anopheles gambiae* mosquito, the analysis of the repertoire of odorant receptors by RNA-seq revealed a specialization of these receptors to the human host of this species [RIN 13].

In ants, comparative transcriptomic approaches and RNA-seq on target species have revealed the role that these olfactory binding proteins (Odorant Binding Proteins and Chemosensory proteins) can play in olfactory recognition, and in particular sex-specific olfactory recognition [ZHO 12]. In addition, the sequencing of entire genomes has revealed the extraordinary diversity of odorant receptors in ant genomes, probably involved in the olfactory recognition functions particularly developed in these insects. In fact, in ants, as well as in other social insects (bees, termites, wasps), the recognition functions of the members of the same colony are based on chemical mediators. The presence or absence of certain mediators, characteristics of the colony of origin, contributes to the tolerance of the "familiar" versus "unfamiliar" triggering agonistic behaviors. These genomic studies in ants have motivated comparative analyses in many species of social insects, allowing for a better understanding of the role of OR and of the origin, evolution and maintenance of sociality in insects. In bees, "omics" have been used to understand the molecular basis of pheromone recognition. It has been shown that the exposition to a brood pheromone induces changes in the expression of hundreds of genes in the brain, with the identification of the *cis* regulatory domains involved in this recognition [ALA 09].

In vertebrates, the genes involved in olfactory reception are distributed in many multigenic families, including OR, expressed in the olfactory epithelium, and the vomeronasal receptors (VR), expressed in the auxiliary olfactory organ called the vomeronasal organ (see Chapter 3). These gene families are very large in most species, exceeding 1,000 OR in the mouse genome, for example, which makes their classification and analysis difficult. The "omics" approaches have helped their identification in a large number of vertebrates. For example, the combination of sequencing of entire genomes and transcriptomes helped to classify the OR and VR in many species of reptiles (Squamata) and mammals. RNA-seq studies on the olfactory epithelium and the vomeronasal organ of mice have also helped to obtain an exhaustive image of OR and VR and revealed the molecular basis of olfactory perception differences between sexes [SHI 12]. Recently, comparative analyses have been conducted on more than 10,000 OR genes of 13 species of placental mammals, revealing that groups of genes have undergone an expansion in certain phylogenetic lineages, African elephants having the largest gene repertoire [NII 14]. Finally, the comparative analysis of human genomes revealed an OR repertoire more restrained in ancient humans (Neanderthals and Denisovan) in comparison to modern humans, suggesting a loss of olfactory function in these lineages [HUG 14]. Recent advances in functional expression also contributed to a more rapid identification of ligands of these OR, which are still only poorly known, particularly in humans.

6.5. From "omics" to physiology: characterizing the modes of production and the modes of reception of active molecules

This involves identifying the organs, tissues and cells, which produce or receive chemical signals, in order to better understand the modes of production, secretion and reception of these signals and their physiological consequences. Thus, in a gastropod (*Biomphalaria*), the combination of transcriptomic and proteomic analyses helped to study the organ responsible for the secretion which allows the immunoprotection of offspring [HAT 10]. The research on chemosensory tissues of mosquitoes by sequencing of the entire transcriptome and analyses of the expression profiles (RNA-seq) is another example of the use of these techniques. In aphids, transcriptomic and proteomic screens have also helped to identify proteins of secretory cells of salivary glands involved in feeding salivary glands involved in feeding on the leaves of the host plant [CAR 11]. Some of these proteins are probably involved in the capacity of aphids to feed on the plants, by yet unknown mechanisms. Some work also helps to study the subcellular level: the search for virulence factors in the Leishmania parasite involved the study of the secretome, which is the secreted portion of the proteome of the animal.

There are also many applications in plants. For example, the first pioneering studies of the transcriptome of aromatic plants producing terpenes, like mint or basil, were carried out after a purification step of the glandular trichomes, which allowed for an enrichment of genes involved in the biosynthesis pathways. In the tomato, the comparison of the transcriptomes of many types of epidermic glands has been carried out to identify the role of each gland in the repelling of insects. In roses, an atlas of the transcriptome of all the organs (leaves, roots, flowers) is available, which helps to understand the spatiotemporal expression of all the genes, some of which are involved in the production of scent. *Catharanthus roseus*, the Madagascar periwinkle, produces hundreds of monoterpene indole alkaloids, some of which have pharmaceutical properties. These molecules probably represent defense compounds. The biosynthesis of these compounds is highly regulated and subdivided, taking place in multiple cells of the leaf (epidermic, laticiferous cells). It is a combination of "omics" approaches, in particular the use of transcriptomes of specific tissues and organs, which helps to reveal the complexity of production of these compounds.

6.6. From "omics" to the role of environment: understanding the impact of biotic and abiotic factors on interactions

One important objective in chemical ecology is also to identify the biotic and abiotic factors which trigger and/or modulate the emission of signals or which influence their reception. Multiple studies concern the expression of genes involved in recognition of interacting partners, in different stress conditions, social conditions, environmental factors, food provisioning and according to different circadian rhythms.

For example, the activation of metabolic genes of a fungus of the *Aspergillus* genus, in response to the presence of 58 bacterial species of the *Streptomyces* genus, was followed by transcriptomic and metabolomic analyses [SCH 09]. In insects, the bee was an emblematic species for the sociogenomic analysis which is an integrative discipline of behavioral biology and which uses the information issued from genomics to study the adaptation and, therefore, the effect of natural selection on gene functioning.

The bee has, therefore, been the subject of all the early studies connecting the influence of the social environment to gene expression. These studies have shown how genetic variations in turn affect the functioning of the brain and modify social behavior [ROB 08]. For example, a study on the brain and ovaries of queen bees helped to highlight the effect of mating on physiological changes and the production of pheromones. Another study on the social wasp, *Polistes metricus*, showed the existence of variations in expressions of dozens of genes involved in chemical communication, under the influence of the maternal and social environments [TOT 07]. By nature, these social interactions which support cohesion in the group require mechanisms of recognition and discrimination of members of the colony as well as the presence of a dialog which, in insects, is mainly chemical.

In plants, transcriptomic analyses of leaves attacked by pathogens or pests, in line with the emission of volatile compounds, are innumerable. Some teams are interested, for example, in the way in which plants react to the perception of volatile compounds emitted by bacteria, during acquired systemic resistance phenomenon. Crop plants have also been the subject of numerous studies aiming to understand the mechanisms responsible for their interactions with pathogens better. Maize, for example, which presents a major agronomic interest for animal and human nutrition and for biofuels, was the subject of intensive research. Researchers working on this plant have integrated databases, including transcriptomic, metabolomic, proteomic and ionomic analyses (mineral analyses) conducted in controlled conditions with various stresses. For example, transcriptomic studies on the biosynthesis genes of Green Leaf Volatiles (GLVs), compounds emitted in response to herbivore attack, highlighted their crucial role as chemical defense signals. These molecules, like Z-3-hexenol, are used by the plant to elicit the defenses of neighboring plants which then respond more rapidly in case of an attack. The genes which are activated during this phenomenon of "priming" are beginning to be characterized by "omic" technologies. We can also cite the pioneering work of the Ian Baldwin group, from the Max Planck Institute of Chemical Ecology (MPI-CE), which has been the first to use the "omics" to study the interactions between plants and herbivores [BAL 06]. The researchers use a species of wild tobacco, *Nicotiana attenuata*, as a model plant, cultivated in the wild to be as close as possible to the natural environment conditions. They conducted numerous experiments with transgenic plants qualified as "mute" (i.e. not emitting at all or partially emitting the signals in an interaction) or "deaf" (i.e. which do not perceive at all or partially a signal emitted by another organism). With this experimental setup, combined with extensive transcriptomic and proteomic studies, they have shown, for example, that the effect of a gene depends on the experimental context in which it is expressed or how a plant can modulate its responses as a function of the different herbivores which attack it. They have also studied the influence of these volatile compounds released by a species of sagebrush, on the expression of the genes and the subsequent rapidity of defense response of neighboring tobacco plants, attacked by caterpillars. This group is also one of the first to have developed multivariate statistical methods for studying the metabolome, in parallel with networks of genes involved in these plant–herbivore interactions. They have in particular applied these methods to the study of the impact of leaf herbivory on the expression of genes in the roots of *Nicotiana attenuata*. Other studies analyzed the modifications of the transcriptome of plants attacked at their root level. In maize, in particular, certain nematodes such as *Diabrotica* are formidable pests, which affect the transcription of numerous root genes [LAW 12]. The defense mechanisms of algae are also beginning to be studied. For example, a transcriptomic analysis in *Laminaria* has shown that oligoguluronates, compounds released by algae following herbivory, lead to a remodeling of transcription activity, with an increase in the expression of genes involved in oxidative stress responses.

Finally, responses to diverse abiotic stresses, such as global warming, also are beginning to be studied in model plant species. In the petunia, a study with transcriptomic experiments shows that emissions of phenylpropanoids diminish following an increase in temperature. This change in emissions is partly due to a reduced expression of some biosynthetic genes [CAN 15]. Such rare studies help to better evaluate the impact that climate change will have on plants, via their chemical interactions with other organisms.

Numerous studies also analyze the impacts of biotic and abiotic factors in other organisms, for example, marine invertebrates. A transcriptomic analysis throughout larval development of the mullosc *Haliotis* (see Figure 6.5) also highlighted endogenous and exogenous chemical factors liable to trigger metamorphosis [WIL 09]. A DNA chip study in the oyster showed that environmental stresses such as pH and temperature interacted in a dynamic manner to affect gene expression [CHA 11].

Figure 6.5. Haliotis asinina *(Boris Laffineur)*

6.7. From "omics" to evolution: understanding and predicting the adaptive value of chemical interactions

One of the great current challenges concerns our understanding of adaptive capacities of natural populations facing environmental changes, whether biotic (e.g. social environment) or abiotic (e.g. climate, trace elements). "Omic" technologies, such as genotyping by sequencing or sequencing of entire transcriptomes or genomes, offer the possibility of analyzing genetic variation among many individuals in a population or comparing genomes of many species and can provide new information on the evolution of genes involved in chemical interactions.

In bees, for example, there are large individual natural differences in the response to pheromones, which help to produce variability in performing ergonomic tasks in the group. These variations are correlated with specific differences in expression of genes at the cerebral level, highlighted by a transcriptomic analysis. In the *Solenopsis invicta* and *Formica selysi* ants, genomic studies resulted in the surprising discovery of the existence of a "social" chromosome. This chromosome contains, according to the authors, the collection of genes indispensable to life in society, with the noticeable presence of several genes involved in communication [PUR 14]. Genomics also facilitated the study of the evolution of receptor families involved in the functions of taste and smell in some insects. In the pea aphid, "omics" technologies have helped to decrypt the genetic bases of the ecological specialization, which involves mechanisms of chemical recognition between the insect and its host plant. Targeted re-sequencing techniques have helped to obtain polymorphism and divergence data for more than 500 candidate genes potentially involved in divergence among races of aphids specialized in different host plants. These studies revealed the role of some odorant and gustatory receptors, whose evolution by variation in the copy number in genomes and by positive selection could be at the origin of the adaptive divergence among different aphid host races [SMA 12]. The use of NGS, and in particular of RNA-seq, was also able to support information on the mechanisms at the origin of the specialization of *Aedes aegypti* mosquitoes to humans [MCB 14]: by comparing genes among "domestic" populations of *Aedes* biting humans and its ancestral form living in forests and preferring to bite non-human animals, this study revealed that the preference for human odor is linked to an increase in the expression and sensitivity of an olfactory receptor, OR4. The evolution of this receptor expression, which recognizes sulcatone, a molecule characteristic of human odor, could, therefore, be at the origin of the specialization of these populations of *Aedes* to the human species.

In vertebrates, comparative studies based on NGS data have helped to address the evolution of receptor families involved in olfaction. For example, the comparison of VR among species of reptiles and mammals helped to better understand how each of these groups was able to adapt to the terrestrial environment, by expansion of specific gene families (type 1 VR in mammals and type 2 VR in reptiles) [BRY 13]. However, there are still very few studies using the "omic" approaches and which are undertaken at the populational scale in vertebrates (see Chapter 2). In the house mouse, current studies combine re-sequencing of entire genomes and transcriptomics (RNA-seq) on olfactory organs (in particular the vomeronasal organ) and are just beginning to reveal the complexity of the genetic bases and the diversity of genetic changes (regulation of gene expressions or protein changes) underlying the evolution of olfactory recognition and sexual isolation in the course of a speciation process.

In plants, few genomes have been entirely sequenced yet. For the species the genome of which is available, data-mining studies on the genes of metabolic biosynthesis pathways have begun to be successful. The re-sequencing of many varieties sometimes helps to explain the evolution of biosynthesis pathways. In rice, the grain's fragrance is one of the most important agronomic characteristics. The study of the gene *BADH1*, which is involved in the genesis of this fragrance, helped to trace its evolution during the process of domestication [KOV 09]. In the tomato, a study showed that the genes involved in some monoterpene and diterpene biosynthesis pathways were localized to the same cluster as the genes coding for the prenyltransferases, which synthesize the precursors on which the terpene synthases act. This result sheds light on how this cluster was able to evolve by duplication and divergence. Despite the absence of complete genomic sequences, studies bearing on the evolution of the genes involved in the biosynthetic pathways have begun to be numerous. The studies were initially conducted on model plants like *Clarkia* or the petunia. In the petunia, many studies focus on the genetic basis of pollination syndromes. In fact, in this plant, certain species are fragrant, presenting a floral morphology with a long tube corolla, and are pollinized by moths. Other species are lacking odors and are pollinized by hummingbirds. QTL analyses, associated with data mining of transcriptome banks available in this plant, helped to show that a small number of loci governed the traits responsible for these pollination syndromes, some of which are linked to fragrance production. NGS techniques now help to extend the analyses of genetic determinism to other species. For example, a study involving transcriptomics focused on the evolution of the capacity to synthesize eugenol in many species of orchids in the *Gymnadenia* genus. They suggest that a minor evolution in a single gene can induce the synthesis of many molecules, leading to a growing complexity of the bouquet emitted by the different species [GUP 14].

6.8. Conclusions and perspectives

Across several examples, we have seen that "omic" approaches find numerous applications and contribute greatly to improving our understanding of chemical ecology. The development of these large-scale molecular analyses and new technologies, in addition to the boom of bioinformatics, has made it possible to make use of the large amounts of readily available data in different fields. This will enable global and more integrative studies helping to have a less reductionist vision of the mechanisms underlying the chemical interactions among living organisms. This progress is at the origin of a new "omics" discipline called post-genomics (or functional genomics), which seeks to integrate the collection of "omic" data in the same schematic representation, with the ultimate goal of reconstructing the complex network of genes and molecules, which intervene to shape the relationships among organisms and their responses to the environment. The field of chemical ecology has to respond to this challenge in the years to come, by pursuing the development and use of "omics" techniques to reveal an integrated image of the often complex mechanisms underlying chemical interactions among living organisms. In parallel, bioinformatic analysis is indispensable for handling large quantities of data coming from these technologies and must be developed to its full potential.

6.9. Bibliography

[ALA 09] ALAUX C., LE CONTE Y., ADAMS H.A. *et al.*, "Regulation of brain gene expression in honey bees by brood pheromone", *Genes, Brain and Behavior*, vol. 8, pp. 309–319, 2009.

[BAL 06] BALDWIN I.T., HALITSCHKE R., PASCHOLD A. *et al.*, "Volatile signaling in plant-plant interactions: 'Talking trees' in the genomics era", *Science*, vol. 311, pp. 812–815, 2006.

[BRI 13] BRISCOE A.D., MACIAS-MUNOZ A., KOZAK K.M. *et al.*, "Female behaviour drives expression and evolution of gustatory receptors in butterflies", *PLoS Genetics*, vol. 9, p. e1003620, 2013.

[BRY 13] BRYKCZYNSKA U., TZIKA A.C., RODRIGUEZ I. *et al.*, "Contrasted evolution of the vomeronasal receptor repertoires in mammals and squamate reptiles", *Genome Biology and Evolution*, vol. 5, pp. 389–401, 2013.

[CAN 15] CNA'ANI A., MÜHLEMANN J.K., RAVID J. *et al.*, "*Petunia hybrida* floral scent production is negatively affected by high-temperature growth conditions", *Plant, Cell & Environment*, vol. 38, pp. 1333–1346, 2015.

[CAR 11] CAROLAN J.C., CARAGEA D., REARDON K.T. *et al.*, "Predicted effector molecules in the salivary secretome of the pea aphid (*Acyrthosiphon pisum*): a dual transcriptomic/proteomic approach", *Journal of Proteome Research*, vol. 10, pp. 1505–1518, 2011.

[CHA 11] CHAPMAN R.W., MANCIA A., BEAL M. *et al.*, "The transcriptomic responses of the eastern oyster, *Crassostrea virginica*, to environmental conditions", *Molecular Ecology*, vol. 20, pp. 1431–1449, 2011.

[CHE 07] CHERTEMPS T., DUPORTETS L., LABEUR C. *et al.*, "A female-biased expressed elongase involved in long-chain hydrocarbon biosynthesis and courtship behavior in *Drosophila melanogaster*", *Proceedings of the National Academy of Sciences of the USA*, vol. 104, pp. 4273–4278, 2007.

[FEN 11] FENG M., SONG F., ALEKU D.W. *et al.*, "Antennal proteome comparison of sexually mature drone and forager honeybees", *Journal of Proteome Research*, vol. 10, pp. 3246–3260, 2011.

[FIO 09] FIORILLI V., CATONI M., MIOZZI L. *et al.*, "Global and cell-type gene expression profiles in tomato plants colonized by an arbuscular mycorrhizal fungus", *The New Phytologist*, vol. 184, pp. 975–987, 2009.

[GAR 11] GARBEVA P., SILBY M.W., RAAIJMAKERS J.M. *et al.*, "Transcriptional and antagonistic responses of *Pseudomonas fluorescens* pf0-1 to phylogenetically different bacterial competitors", *The ISME Journal*, vol. 5, pp. 973–985, 2011.

[GUP 14] GUPTA A.K., SCHAUVINHOLD I., PICHERSKY E. *et al.*, "Eugenol synthase genes in floral scent variation in *Gymnadenia* species", *Functional & Integrative Genomics*, vol. 14, pp. 779–788, 2014.

[HAT 10] HATHAWAY J.M., ADEMA C.M., STOUT B.A. *et al.*, "Identification of protein components of egg masses indicates parental investment in immunoprotection of offspring by *Biomphalaria glabrata* (*Gastropoda, Mollusca*)", *Developmental and Comparative Immunology*, vol. 34, pp. 425–435, 2010.

[HUG 14] HUGHES G.M., TEELING E.C., HIGGINS D.G., "Loss of olfactory receptor function in hominin evolution", *PLoS One*, vol. 9, p. e84714, 2014.

[KAM 13] KAMPA A., GAGUNASHVILI A.N., GULDER T.A. *et al.*, "Metagenomic natural product discovery in lichen provides evidence for a family of biosynthetic pathways in diverse symbioses", *Proceedings of the National Academy of Sciences of the USA*, vol. 110, pp. E3129–3137, 2013.

[KOR 09] KORB J., WEIL T., HOFFMANN K., FOSTER K.R. *et al.*, "A gene necessary for reproductive suppression in termites", *Science*, vol. 324, p. 758, 2009.

[KOV 09] KOVACH M.J., CALINGACION M.N., FITZGERALD M.A. *et al.*, "The origin and evolution of fragrance in rice (*Oryza sativa* L.)", *Proceedings of the National Academy of Sciences of the USA*, vol. 106, pp. 14444–14449, 2009.

[LAM 15] LAMDAN N.-L., SHALABY S., ZIV T. *et al.*, "Secretome of the biocontrol fungus *Trichoderma virens* co-cultured with maize roots: role in induced systemic resistance", *Molecular & Cellular Proteomics*, vol. 14, no. 4, pp. 1054–1063, 2015.

[LAW 12] LAWRENCE S.D., NOVAK N.G., KAYAL W.E. *et al.*, "Root herbivory: molecular analysis of the maize transcriptome upon infestation by southern corn rootworm, *Diabrotica undecimpunctata howardi*", *Physiologia Plantarum*, vol. 144, pp. 303–319, 2012.

[LEG 11] LEGEAI F., MALPEL S., MONTAGNE N. *et al.*, "An expressed sequence tag collection from the male antennae of the noctuid moth *Spodoptera littoralis*: a resource for olfactory and pheromone detection research", *BMC Genomics*, vol. 12, p. 86, 2011.

[MCB 14] MCBRIDE C.S., BAIER F., OMONDI A.B. *et al.*, "Evolution of mosquito preference for humans linked to an odorant receptor", *Nature*, vol. 515, pp. 222–227, 2014.

[NII 14] NIIMURA Y., MATSUI A., TOUHARA K., "Extreme expansion of the olfactory receptor gene repertoire in African elephants and evolutionary dynamics of orthologous gene groups in 13 placental mammals", *Genome Research*, vol. 24, pp. 1485–1496, 2014.

[PIE 02] PIEL J., "A polyketide synthase-peptide synthetase gene cluster from an uncultured bacterial symbiont of *Paederus* beetles", *Proceedings of the National Academy of Sciences of the USA*, vol. 99, pp. 14002–14007, 2002.

[PUR 14] PURCELL J., BRELSFORD A., WURM Y. *et al.*, "Convergent genetic architecture underlies social organization in ants", *Current Biology*, vol. 24, pp. 2728–2732, 2014.

[RIN 13] RINKER, D.C., ZHOU, X., PITTS, R.J. *et al.*, "Antennal transcriptome profiles of anopheline mosquitoes reveal human host olfactory specialization in *Anopheles gambiae*", *BMC Genomics*, vol. 14, p. 749, 2013.

[ROB 08] ROBINSON G.E., FERNALD R.D., CLAYTON D.F., "Genes social behavior", *Science*, vol. 322, pp. 896–900, 2008.

[RUN 06] RUNYON J.B., MESCHER M.C., DE MORAES C.M., "Volatile chemical cues guide host location and host selection by parasitic plants", *Science*, vol. 313, pp. 1964–1967, 2006.

[SCA 08] SCALLIET G., PIOLA F., DOUADY C.J. *et al.*, "Scent evolution in Chinese roses", *Proceedings of the National Academy of Sciences of the USA*, vol. 105, pp. 5927–5932, 2008.

[SCH 09] SCHROECKH V., SCHERLACH K., NÜTZMANN H.-W. *et al.*, "Intimate bacterial–fungal interaction triggers biosynthesis of archetypal polyketides in *Aspergillus nidulans*", *Proceedings of the National Academy of Sciences of the USA*, vol. 106, pp. 14558–14563, 2009.

[SHI 12] SHIAO M.S., CHANG A.Y., LIAO B.Y. *et al.*, "Transcriptomes of mouse olfactory epithelium reveal sexual differences in odorant detection", *Genome Biology and Evolution*, vol. 4, pp. 703–712, 2012.

[SMA 12] SMADJA C.M., CANBÄCK B., VITALIS R. *et al.*, "Large-scale candidate gene scan reveals the role of chemoreceptor genes in host plant specialisation and speciation in the pea aphid", *Evolution*, vol. 66, pp. 2723–2738, 2012.

[SPL 14] SPLIVALLO R., DEVEAU A., VALDEZ N. *et al.*, "Bacteria associated with truffle-fruiting bodies contribute to truffle aroma", *Environmental Microbiology*, vol. 17, pp. 2647–2660, 2015.

[TOT 07] TOTH A.L., VARALA K., NEWMAN T.C. *et al.*, "Wasp gene expression supports an evolutionary link between maternal behavior and eusociality", *Science*, vol. 318, pp. 441–444, 2007.

[VER 05] VERDONK J.C., HARING M.A., VAN TUNEN A.J. *et al.*, "*Odorant1* regulates fragrance biosynthesis in petunia flowers", *The Plant Cell*, vol. 17, pp. 1612–1624, 2005.

[VOG 10] VOGEL H., HEIDEL A.J., HECKEL D.G. *et al.*, "Transcriptome analysis of the sex pheromone gland of the noctuid moth *Heliothis virescens*", *BMC Genomics*, vol. 11, p. 29, 2010.

[WIL 09] WILLIAMS E.A., DEGNAN B.M., GUNTER H. *et al.*, "Widespread transcriptional changes pre-empt the critical pelagic–benthic transition in the vetigastropod Haliotis asinina", *Molecular Ecology*, vol. 18, pp. 1006–1025, 2009.

[ZHO 12] ZHOU X., SLONE J.D., ROKAS A. *et al.*, "Phylogenetic and transcriptomic analysis of chemosensory receptors in a pair of divergent ant species reveals sex-specific signatures of odor coding", *PLoS Genetics*, vol. 8, p. e1002930, 2012.

7

Metabolomic Contributions to Chemical Ecology

Recent developments in metabolomics enable new approaches to characterize the chemically mediated interactions between organisms and their environment. Based on the rapid assessment of the global content of small metabolites in a living organism, metabolomic techniques complement the traditional methods of bioassay-guided fractionation to identify chemical cues and to elucidate the impacts of chemical communication in ecosystem functioning.

7.1. Definition of metabolomics

In the same way as for other "Omics" sciences, metabolomics is a global approach which aims at the analysis without *a priori* knowledge of the largest possible number of metabolites of low molecular weight (<1,500 Da or amu) present in a cell, a tissue, a biological fluid or an organism placed in given conditions. Metabolomics is the most recent of the "Omics" sciences and it was not until 1998 that the term "metabolome" was proposed by Fiehn *et al.* [FIE 02]. One of the particularities of metabolomics by comparison to other "Omics" is that, taking into account the diversity of the physicochemical properties of the small molecules produced by organisms, an exhaustive coverage of the metabolome remains impossible by the current analytical techniques.

Chapter written by Philippe POTIN, Florence NICOLÈ and Olivier P. THOMAS.

Whatever their properties, the metabolites, having originated in specific biosynthesis pathways, play key roles in interactions between organisms and their environment, and their regulation contributes important information in numerous disciplines.

The field of application for metabolomics is quite vast and includes diagnosis in medical sciences, nutrition in the agri-food sector or for phenotyping, and systematics for classifying and/or characterizing different types of strains (bacteria, fungi) and organisms (plants, insects, etc.). In ecology, a major asset of metabolomic approaches is that they are applicable to non-model organisms. Thus, numerous applications are possible, for example, 1) in ecophysiology, to substantiate the biochemical responses of organisms to biotic and abiotic conditions of their environment; 2) in ecotoxicology to indicate the potential risks of chemical pollutants on organisms in their environment; and especially 3) in chemical ecology to contribute to the identification of chemical mediators (semiochemical or bioactive compounds) involved in ecological interactions.

The topics that involve metabolomic approaches in chemical ecology vary greatly and their interests are illustrated in many examples cited in other chapters of this book. After a description of the strategies of metabolomic approaches and of their specificities for chemical ecology, this chapter details several examples of such approaches to resolve questions in chemotaxonomy, to decipher the pathways of biosynthesis of chemical mediators, and also in studies of biotic interactions. It also higlights a booming domain of environmental metabolomics which explain the interactions between organisms and their environment to address the functioning of an organism at the molecular level and/or inform on the state of the environment.

7.2. Different strategies of the metabolomic approaches

Different levels of analyses can be envisioned for identifying qualitative or quantitative variations of chemical compounds among different biological samples (Figure 7.1): targeted or non-targeted approach, metabolite fingerprinting and metabolite profiling. The targeted approach focuses on one or several molecules of interest while the non-targeted analysis is interested in the collection of data for all observed metabolites. Metabolite fingerprinting helps evaluate differences in the biological samples by the presence or absence of molecules quickly and requires neither identification nor quantification [FIE 02]. This rapid technique, exhaustive and non-targeted, is appropriate for obtaining a global image of the sample and achieving profile recognition. It also helps to identify the discriminating regions of chromatograms and can lead to more comprehensive targeted analyses. Metabolite profiling requires that the molecules be identified and quantified. It can be non-targeted, or focused on a class of compounds or a specific metabolic pathway.

	Advantages	Pitfalls
Metabolite fingerprinting	Global coverage of metabolome Pattern recognition High-throughput screening	No compound identification Limited statistical analyses for binary data
Metabolite profiling	Global coverage of metabolome Quantitative Compound identification	Large amount of data (problem to handle data, computer limitations) Semi-quantitative Medium-throughput screening
Targeted analysis	Quantitative Compound identification Low detection limit	Low throughput screening No detection of untargeted compounds Expensive Need for pure standards

Figure 7.1. *Advantages and disadvantages of different metabolomic approaches*

7.3. The different steps for conducting a metabolomic study

Metabolomic approaches require various capabilities, from experimental design to data integration (Figure 7.2), from analytical chemistry to informatics; from biochemistry to complex statistical analyses; and all of this, eventually, integrated in a larger corpus of omic data (see Chapter 6).

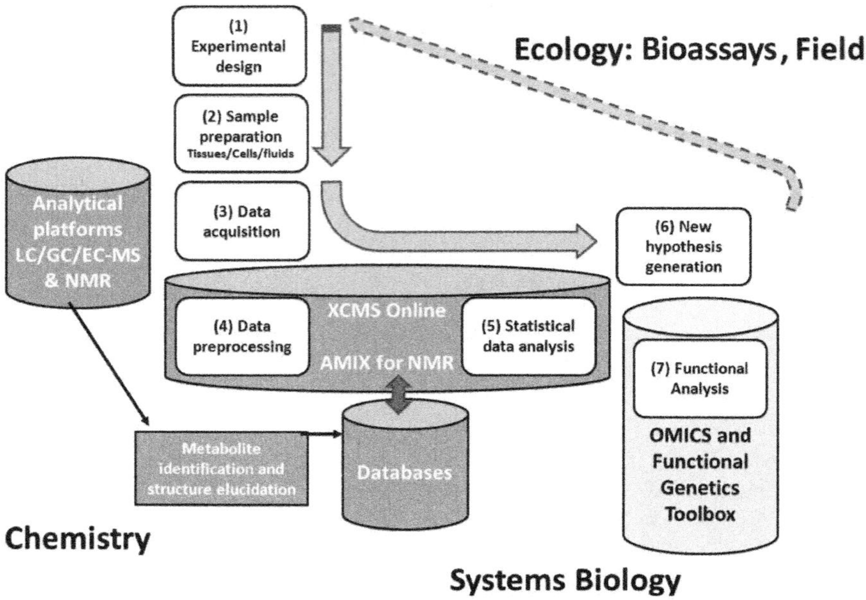

Figure 7.2. *Metabolomics workflow. The different steps from experimental design to the use of metabolomic data in relation with the analytical platforms, bioinformatics, databases and the interdisciplinarity of approaches*

7.3.1. *Experimental design and sampling*

Experimental design (or design of experiments DOE, Figure 7.2) consists of establishing a plan of the experiment, before its implementation, while taking into account the different factors that have an influence on the problems, while minimizing the influence of technical and methodological variability. An optimal experimental design helps extract a maximum of information linked to the problem while integrating the other sources of variability and maximizes the cost efficiency.

In chemical ecology, the experimental design is critical because the sources of variability are multiple. Certain sources of variability can be experimentally manipulated and controlled (we speak of fixed factors): variations in physicochemical or environmental conditions, for example. Others cannot be controlled by the experimenter and all the modalities of the factor are not present in the experiment; they are chosen in a random manner (we speak of random factors). The experimenter is not interested in each individual modality but in the global effect of the factor: genotype/individual, geographic locations, years, etc. The experimental design must also integrate the relations among the factors (crossing or interlocking). Two factors are called crossing if each modality of one appears with each modality of the other (A x B). A factor A is embedded in a factor B if each modality of A is associated with exactly one modality of B (A[B]).

The variability of metabolic profiles can also depend on bias and technical and methodological artifacts. It is, therefore, essential to quantify this source of variability in operating tests of repeatability (application of the same measurement conditions) and reproducibility (varying the measurement conditions). This type of analysis must be generalized to each metabolomic experiment to evaluate the reliability of the results and must appear in the scientific publications. Second, the results of these analyses must be taken into account in the experimental design to minimize the exogenous sources of variability and favor the "informative" sources (for a review dedicated to these metabolomic questions, see Hendriks *et al.* [HEN 11]).

The number, type and relations among the factors, as well as the technical and methodological variabilities, help to define the number of replicates by modality. This number of replicates must be optimized as a function of the budget and availability of samples. From a statistical point of view, the sample size must be optimized by power assessments. Often ignored, the power assessments are a unique tool for determining the sampling effort necessary to highlight the role of ecological factors in the variations of chemical profiles. The power assessments are well established for univariate analyses, but the software is often limited when faced with complex metabolomic multivariate data (see meanwhile, MetSizeR, Nyamundanda *et al.* [NYA 13]). We can evoke the possibility of power calculations on a multivariate analysis of variance (MANOVA) under the hypothesis that the multivariate model is linear and that the errors are Gaussian. For abundant data, this analysis helps test the significance of chemical profiles among groups. A permutational alternative of MANOVA helps to overcome the normality of errors.

For a sampling strategy, the collection and preservation of samples are also two key steps to minimize the sources of artifacts or loss of information. Notably, the quenching of metabolic reactions is crucial, either by a rapid freezing of collected material or by the addition of a solvent which inhibits any reaction and also preserves the samples from bacterial contaminants. Further manipulation of the samples and the choice of extractive methods appropriate to the types of biomass are then applied to pursue these objectives.

7.3.2. *Analytical approaches*

To implement an experimental design in chemical ecology, suitable for metabolomic analyses, the two major analytical methods currently used are mass spectrometry (MS) and nuclear magnetic resonance (NMR), the choice of which first depends on the conditions of metabolite extraction and will then depend on the conditions used for separation ionization, and also on the solubility of the chemical constituents. The bases and principles of these widely used methods in chemical ecology are presented in Chapter 8. The technological progress in this domain is considerable [ZHA 12].

7.3.3. *Data processing*

Informatics tools for data processing become indispensable for making sense of the large amount of data generated by metabolomics approaches. They are simultaneously essential for the pre- and post-data processing phases [GOO 04]. The pre-processing consists of working on the raw data coming from the analyzer with the goal of obtaining finalized data matrices. These matrices can be in the form of binary data (presence/absence of peaks: fingerprinting) or the relative or absolute abundance (profiling) or as a percentage of the total. In the post-processing phase, these matrices are statistically analyzed to identify the most biologically significant metabolites. As a function of the type of data and the problem of the research, the data processing must be adapted; it can be univariate or multivariate, targeted or non-targeted, supervised or non-supervised.

7.3.3.1. *Pre-processing: processing of raw data*

Abundant literature exists on this phase of pre-processing which aims to extract, transform and compile the chromatographic and/or spectrometric data of each sample in the form of multivariant statistics, with molecules in columns and samples in rows. This work can prove particularly difficult and time-consuming if the number of analyses and metabolites is important. Numerous bioinformatic tools have been developed to achieve one or more pre-processing functions and choosing the most adapted tool for the data and the problem can prove complex (see Chapter 8). These tools help reduce the processing time from the acquisition of raw data to the identification of compounds efficiently. Certain tools are free (XCMS or Metalign). Others use commercial software such as MatLab. Still others use commercial software developed by machine manufacturers (Agilent's genespringMS, Water's MarkerLynx, ABSciex's MarkerView, ThermoFisher's Sieve, Bruker's Amix). New software is regularly created and reviews are regularly published for compiling these dozens of software solutions. In this complex imbroglio, procedures have been undertaken to standardize the exchange formats of the data and the analyses processes. The work groups are in place to envision common procedures, processing tools and the means of managing the handling of increasingly large and complex metabolomic data. These procedures must help to reinforce the effectiveness of metabolomic research and greatly facilitate exchanges and collaborations among laboratories. Thus, to extract raw data, it is preferable to use the free data exchange formats netCDF and mzXML. They allow manipulating the data with most types of bioinformatic tools.

In this vision of workflow standardization, integrated data processing programs emerge: XCMS, MetAlign, MZmine2, MetaboAnalyst, metaP – server, MetAtt and MSEasy. For example, to analyze chemical composition data coming from GC–MS, MSEasy software [NIC 12] was created to process large quantities of data (https://sites.google.com/site/rpackagemseasy/, http://cran.rproject.org/web/packages/MSeasy/index.html). For chemical ecology, it opens the field of evolutionary and populational studies on volatile compounds because these require the analysis of numerous and often unknown samples. The different bioinformatic tools offer data processing pipelines simultaneously including the steps of pre- and post-processing. Some offer data processing on more powerful external servers; integrating file download functionality by streaming directly from the machine to the server which converts and directly processes the files (XCMS online). They manage the classic steps of alignment of retention time, normalization, and identification, as well as statistical processing aiming to identify compounds of interest. This step is covered in more depth in the post-processing section. First developed in France [GIA 14], this platform has been introduced internationally and is from now accessible by a web interface in a Galaxy environment which offers all its entire functionality by integrating all of its statistical and database query tools in the same analyses suite (www.http://galaxy.workflow4metabolomics.org/).

7.3.3.2. *Post-processing: statistical processing of finalized matrices*

Once the data clean-up has been completed during the pre-processing phase, the finalized matrix, with the samples in rows and the metabolites in columns, can be obtained. The “metabolite” variables can be coded in different ways as a function of the extraction methods used and the problems: binary coding in presence/absence, relative or absolute abundance, quantity in percentage of the total. In the latter case, the “metabolite” variables are dependent on one another. When the data are coded in abundance, it is possible to carry out logarithmic type transformations, centering and reduction, to improve the implementation of the statistical analyses.

Appropriate analyses should be selected for each type of data (Figure 7.3). We can first differentiate between the univariate and multivariate approaches, which are used to address distinct problems. The univariate analyses will help to analyze each compound independently. They help test whether the emission of a different metabolite is a function of a given factor (the t-test or the Wilcoxon test, if two modalities to compare; ANOVA or the Kruskal–Wallis test for more than two modalities; the Wilcoxon test and the Kruskal–Wallis test are the non-parametric alternatives that do not require the normality of residues and equality of variances between modalities). Then, we must envision as many tests as the number of metabolites, but this approach helps to identify biomarkers with a significant p-value. They constitute a tool of choice in targeted approaches on one or a few molecules.

By contrast, multivariate analyses help to process a chemical profile in its entirety. If the user wishes to compare the profiles of chemical compounds, a multivariate analysis will bring an objective response. Beyond the global comparison of the profile, multivariate analyses also help show associations among metabolite peaks, useful for inferring the pathways of biosynthesis. Finally, multivariate analyses help visualize the distribution of the different components of the variability of chemical profiles (intra-group versus inter-group variability). They can apply to all types of data (binary, abundance or percentage) and on targeted or non-targeted studies. In the case of non-targeted and exhaustive studies, they first help visualization in their entirety, followed by a simplification by suppressing the less structuring variables.

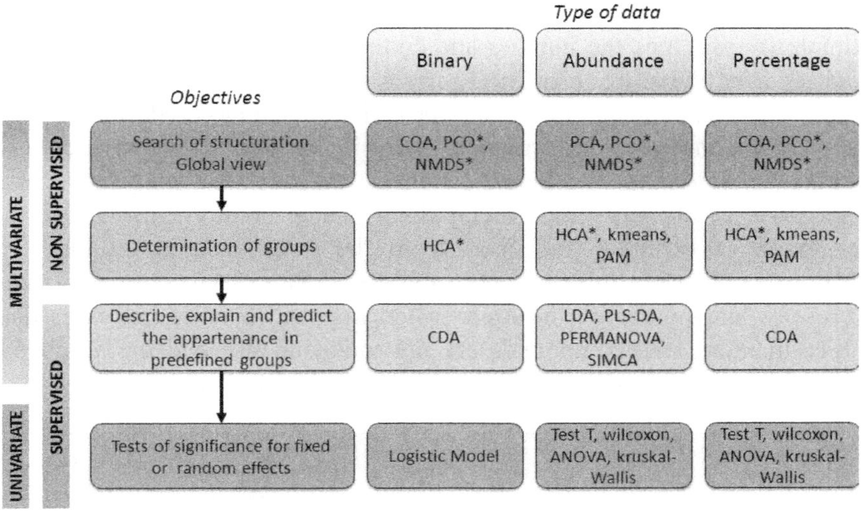

* These analyses are based on the calculation of a distance matrix and the results may differ in function of the distance choosen.

Figure 7.3. *The different types of analyses in metabolomics as a function of data type and study objective*

The first step in the processing of chemical profiles consists of carrying out data visualization with a descriptive non-supervised multivariate analysis (i.e. No known groups or groups not informed in the analysis): principal component analysis (PCA, Figure 7.4) for the data in abundance, correspondence analysis for binary data and percentages, and principal coordinates analyses (PCO) and non-metric multidimensional scaling (NMDS) for all types of data via the calculation of a distance matrix among samples. However, the reduction of the chemical profile to a distance often leads the PCO and NMDS to be less informative than the other two methods. The NMDS method can prove fairly ineffective for structuring the data because it is a non-parametric and non-quantitative technique of ordination, which is only interesting at the global level of similarities among individuals.

Second, identification of a number of groups with different chemical profiles will require different clustering methods: hierarchic clustering in particular ascending or agglomerative, or analyses by the flat or k-means partitioning clustering or dynamic clustering method (partition around medoid). Indices (e.g. Silhouette or BIC) help to identify the number of optimal groups. Once the number and composition of the groups have been obtained, it is, therefore, possible to use a supervised approach, which will seek to describe, explain and predict the affiliation data to predefined groups. Only correspondence discriminant analysis applies to binary data or percentages. For data in abundance, linear discriminant analysis requires application conditions that metabolomic data rarely fulfill (normal distribution of variables and equality of the covariance in groups). The partial least squares discriminant analyses (PLS-DA) overcome these restrictions and overcome the limits often imposed by metabolomics data: multicollinearity (the compounds are not independent), and the number of variables is greater than the number of samples (Figure 7.4).

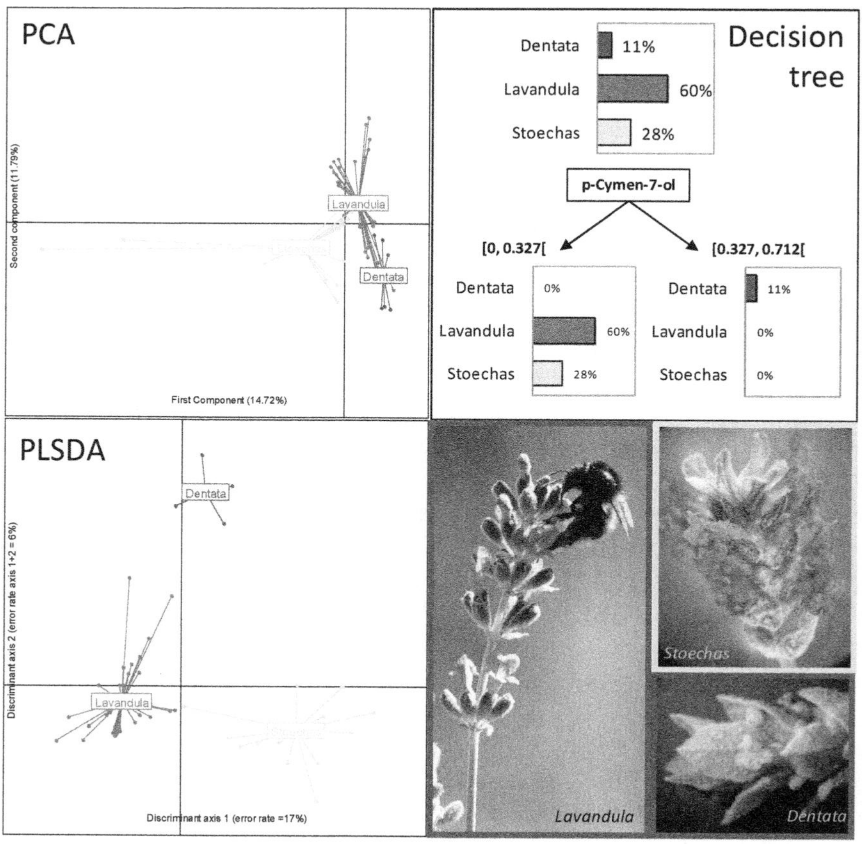

Figure 7.4. *Different post-processing analyses performed on the composition of terpenes of three sections of the genus Lavandula*

For the PCA and PLS-DA, "sparse" analyses perform a selection from automatic variables. More recently, more complex methods of automatic learning from data mining have been applied to metabolomic data. Decision trees aid the automatic selection of discriminant variables, supply a simple representation of the decision model (the tree) and constitute an exploratory technique to understand complex metabolic profiles. The artificial neuron network was successfully used to classify chemical profiles and is becoming one of the most popular methods for understanding patterns. Data visualization and interactivity are now used to visualize metabolomic data in order to facilitate the interpretation of complex data-sets. XCMS online [GOW 14] offers cloud-plots, PCA and interactive heatmaps (i.e. the heatmaps are graphical representations of correlation matrices). These two types of visualization help the user personalize the display and easily select the most interesting compounds.

7.3.3.3. *Identification of metabolites*

The compounds identification step often proves to be most challenging. In addition, in an exhaustive and non-targeted approach, if the number of compounds is large, it may be better to focus on the identification of targeted compounds of interest in the post-processing phase (see also Chapter 8).

In GC or LC/MS, the identification of a compound is carried out on the basis of its fragmentation spectrum and its retention time (or index of retention based on a range of linear calibration by hydrocarbons). Database querying software compares the mass spectrum of the peak with the available mass spectrum in the database to identify compounds. This querying, therefore, constitutes a first simple approach, fast and automated, for identifying the structure of molecules. The identification depends, therefore, on the richness and quality of the databases. Some researchers think that one of the current limits of metabolomics is its dependence with regard to databases that are not yet complete. The querying software sends the assignments of molecules, even if the unknown molecule is not in the database. These assignments are based on the similarity of fragmentation spectra and different coefficients of identification quality can be offered. Thus, a good coefficient is not sufficient for naming a molecule with reliability. Automatic identifications must be confirmed by injection of standards (reference molecules) on the same analytical support. The standard must, therefore, present the same retention time, the same exact mass and the same fragmentation profile as the unknown compound.

Meanwhile, if the databases do not give a satisfactory response, identification of a molecule from MS data is complex and is often a source of errors. In fact, for numerous non-model organisms and in particular eukaryote marine organisms, but also fungi, plants or other non-model organisms, the compounds present in the databases are still very limited. The structural elucidation procedure (nitrogen rule, use of isotope mass, fragmentation, etc.) accessible from high-resolution MS data, therefore, become very restrictive. In these conditions, the classic methods of chemical fractionation, purification of metabolites by preparative or semi-preparative chromatography (see Chapter 8) often prove indispensable for constituting a first database of metabolites from non-model organisms. This part requires successive steps of fractioning and purification by chromatography for preparative HPLC quantities of extracts to purify greater than 100 mg approx., semi-preparative if they comprise between 10 and 100 mg and analytical if they comprise between 1 and 10 mg) to help the structural identification by NMR of pure compounds initially detected by MS. It is good to recall here that, even if the NMR techniques become increasingly sensitive, quantities of pure compounds greater than 0.5 mg or even 1 mg are required for identifying unequivocally an organic molecule. These chemical approaches of natural products, often time consuming, can sometimes be accelerated by qualified metabolomic methods of dereplications and targeting of purifications in instances where the chemical mediators are more abundant.

Recent initiatives have been introduced, in particular for metabolomic experiments for plants such as Armet but also the Metabolomics Standard Initiative [FIE 07]. These initiatives help, among others, to share a certain number of data concerning the standards and thus facilitate by automation the annotation of compounds. Much software today helps explore the databases to propose chemical structures. Most of the time they use the 1D and 2D NMR data, such as the Chenomx, Amix (Bruker), Metabominer, rNMR or even BMRB software.

Whatever the approach, the work of identification of the peaks remains a key and obligatory step in the goal of researching biomarkers. The bioinformatic tools can help group similar molecules but human intervention will always be necessary for certain identification of molecules which thus help to build solid databases.

7.3.3.4. *Integration of data into metabolic networks and systems biology*

Metabolomic studies must maximize the potential of the available information. Rather than only furnishing a list of metabolites with differential expressions in the data, they must aim to identify the functional roles and the mode of action of metabolites. New algorithms such as mapping and neuronal networks help to access a higher level of integration. Metabolomic "metadatabases", which integrate and link different levels of information such as METLIN, contribute to identifying the functional roles of metabolites. Bioinformatic tools can help identify the biosynthesis pathway through which a metabolite is derived. The researcher can use this understanding of the biology of the compound to propose a mechanism of action.

7.4. Applications of metabolomics

7.4.1. *Chemical biodiversity and chemotaxonomy*

Metabolomic approaches can constitute a first step for classifying chemical biodiversity when it involves non-model or little known organisms. With the same tools as genomics and population genetics, it is possible to analyze the structure of chemical diversity at different taxonomical or hierarchical levels (castes, ranks, varieties, populations, species, genera, etc.). A structure can also be researched at different spatial (locations, bioclimatic regions, ecotypes) or temporal (phenology, thermogenesis, season, annuality) scales. Similar to genetics, chemical screening can aid in the resolution of the complex of species, for example, in the case of the *Parazoanthus axinellae* species in the marine Zoantharia [CAC 15]. To obtain this global image of the chemical diversity of an organism and understand the different sources of variability very often constitute a prerequisite to the most targeted studies (see Chapter 1).

The diversity of chemical compounds has been used directly for centuries for direct classification. Initially, these classifications grouped organisms by their use or their danger to humans and said nothing of phylogenetic or evolutionary relationships [WIN 99]. The use of secondary metabolites as taxonomic characteristics developed with the advance of chromatography, giving birth to a research discipline named chemotaxonomy. To undertake taxonomy from secondary metabolites also poses difficulties: secondary metabolites do not constitute independent characters since some can be issued from the same precursor or be regulated by the same genes. To alleviate this difficulty, it is possible to use as a taxonomic characteristic not only the separate compounds but also the known biosynthetic pathways, or to regroup the compounds by structural type. This type of coding of chemical data can lend an evolutionary direction to classifications. Additionally, secondary metabolites have an essential role in survival (communication, defense, pollination, etc.), the expression of secondary metabolites of a given structural type sometimes appears in different taxonomic groups and reflects the adaptations and strategies from the history of life rather than evolutionary relationships. Thus, the variability of metabolic expression makes secondary metabolites unreliable taxonomic markers.

7.4.2. *Study of the regulation and evolution of metabolic/ biosynthesis pathways*

The metabolomics approach is becoming an essential tool in the study of metabolic pathways leading to secondary metabolites [SUM 15], or to "specialized" metabolites, involved in chemical communication among living organisms. Despite the sometimes spectacular complexity of molecular architectures provided by the so-called secondary metabolism, the number of chemical precursors and the biochemical reactions put into play are relatively limited. Gene coding for key enzymes catalyzing the formation of these products is the result of a long evolutionary process, but the role of biosynthetic enzymes for which they encode has been hardly elucidated before the advancement of the first of the omic sciences, genomics, and then through confrontation with the biochemical data on these native or recombinant enzymes and the combination of other omics approaches (see Chapter 6). The different approaches put in play to help understand these biosynthesis pathways can be summarized as below (see Figure 7.5):

– As a first step, the first experimental studies of understanding of biochemical pathways leading to secondary metabolites have all been achieved through "feeding" experiments by using potential precursors marked isotopically (radioactive as ^{3}H or ^{14}C with a very sensitive detection, or stable such as ^{2}H, ^{13}C or ^{18}O detected in a less sensitive manner by MS or NMR). The detection of a marked metabolite after an *in vivo* metabolization step helps to highlight the incorporation of this precursor in the described metabolic pathway and, therefore, infer its key roles. Even if, today, these experiments are used less frequently in view of the greater understanding obtained by genomics, it is worth noting that they are nonetheless indispensable in the case of little or poor biosynthesis knowledge, in particular for the products of complex and original structures such as those produced by marine invertebrates.

– These preliminary *in vivo* experiments then lead to proposals of putative metabolic pathways for the production of substances of interest. These hypotheses, therefore, help to mine the genes coding for known enzymes in the genome of the organism, and thus confirm the involved reactions. These studies thus help an eventual localization in the genome of genes or of a group of genes involved in the relevant biosynthesis (Figure 7.5)

– A confirmation must finally be given by the *in vivo* studies. Enzymes can be produced under recombinant form and an *in vitro* study takes place in the presence of these enzymes to validate certain reactions from each of the putative intermediates. Studies in molecular biology help insert the gene or a cluster of genes into a heterologous host to monitor the production of the target molecule in the environment and provide the formal proof of biosynthesis.

In this framework, the combination of metabolomic approaches to genomic, transcriptomic or proteomic analyses helps accelerate the comprehension of metabolic pathways and also their regulations. These methods are well known today for particular families of natural products often produced by clusters of genes, which are beginning to be well described [GEN 12]. Thus, the application of combined approaches for the polyketides or peptides produced by microorganisms responds to numerous questions. For example, metabolic networks linked to Colibactin were able to be described by targeted metabolomic approaches on cultures of *Escherichia coli* [VIZ 14]. These metabolomic approaches thus have very important applications in biotechnology. The study combined numerous Actinobacteria, thus leading to the highlighting of the potential of these microorganisms for the production of bioactive natural substances [DOR 14]. These approaches have also helped the optimization of production conditions of natural high-value-added molecules such as taxanes [CUS 14].

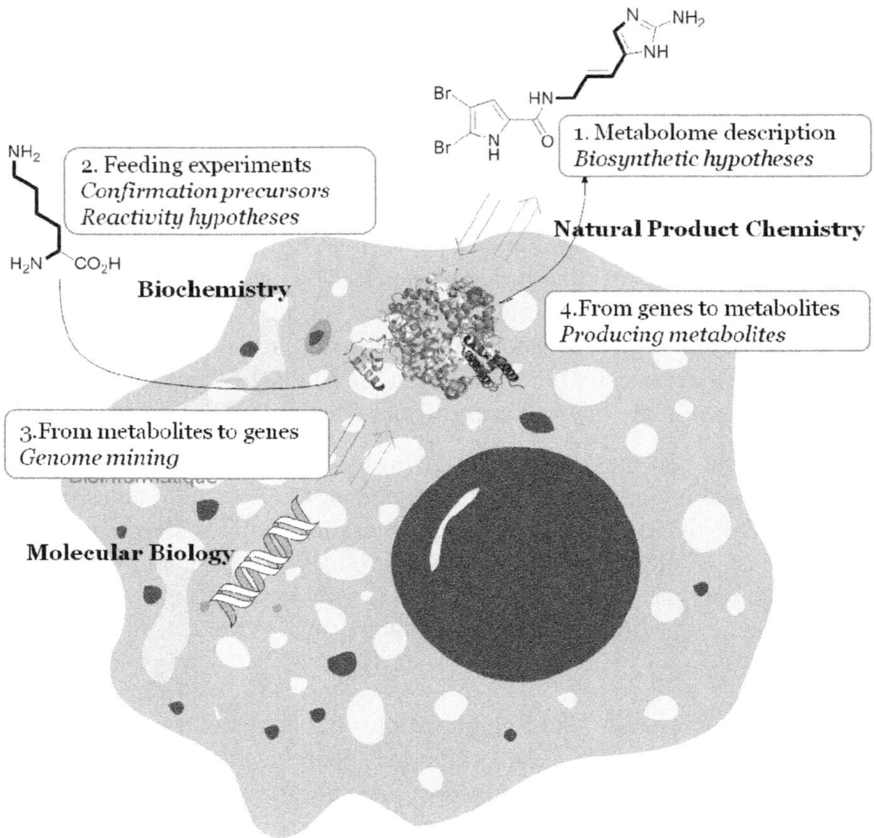

Figure 7.5. *The elucidation steps of new biosynthesis pathways from the description of the metabolome (1), experiments using marked precursors (2), the incorporation monitoring which helps send the metabolite to the genes (3) and then carry out the final steps of functional validation (4)*

More recently, with the development of sequencing technologies and powerful phylogenetic analysis software (PAUP, phylip, MEGA, phyML, MrBayes, etc.) (see Chapters 6 and 8), phylogenetic reconstruction based on DNA sequencing was developed. These molecular phylogenies constitute a useful framework for studying the biosynthesis pathways in the light of evolution. It is possible to superimpose the presence of secondary metabolites on phylogenetic trees by tools such as MacClade or AdePhylo.

This approach helps to identify compounds or biosynthesis pathways, which constitute apomorphies (characters derived from a common ancestor and present in all members of a monophyletic clade) or homoplasies (evolutionary convergence developed independently in different taxonomic groups or non-inherited characters of a common ancestor, defining a polyphyletic group).

7.4.3. *Contributions to functional ecology*

In the environment, chemical signals play essential roles at different ecological levels, from the species (e.g. lifecycle dynamic, larval development, phenology, morphogenesis, growth vs. defense), to the trophic network (e.g. herbivory, "brown food chain", prey/predator relation) and to the structuring of communities (allelopathy), and also at the level of biogeochemical cycles integrating the microbial compartment, whether it is in a marine or a terrestrial environment.

Functional ecology is a field of application of metabolomics. It generally accompanies an "Omics" type of combined approaches and is made possible by a certain understanding of chemical and molecular processes driving the physiology of a species and its biotic interactions. In fact, the change of scale linked to the transfer of studies from the laboratory to the natural environment induces different levels of complexity, such as:

– the impact of abiotic factors and ontogeny on the synthesis and liberation of secondary metabolites and on biotic interactions, whether trophic, symbiotic, endophyte, pathogens or others;

– the great dilution or spatiotemporal transformation of chemical signals; and

– the necessity to integrate the intra- and interspecific communication network, at the community or ecosystem scale.

Thus, this domain of research, one of those most advanced in terms of understanding mechanisms of chemical signaling among Eukaryotes, has evolved in recent years to more complex studies at the community scale, notably through the recent developments of metabolomic approaches coupled with reverse genetics on a model such as the plant *Nicotiana benthamiana* [GAQ 14]. Likewise, the contribution of metabolomics to metagenomics has helped explore the dynamics and functioning of certain bacterial communities (soil, digestive tract) and to progress in the comprehension of bacterial assemblages (biofilms). In marine environments, biotic interactions governed by chemical communication seem to drive the dynamics of certain planktonic populations but often the nature of the chemical signals and the mechanisms responsible for their biosynthesis remain unknown. Recent advances concerning allelopathic interactions occurring in algal blooms involving Diatoms and the dinoflagellate predator *Karenia brevis* have recently been published [POU 14] and they simultaneously combine global metabolomics and proteomics to reveal the impact of chemical mediation on the competition. In the terrestrial environment, the role of secondary metabolites in the ecology of interactions and more particularly in biodiversity via allelopathic processes remains yet to be explored. In these two cases, metabolomics will help identify markers of this ecological process.

The study of chemical determinants of the lifecycle (larval development, morphogenesis, growth phase) has largely been developed in social insects and certain animals. Exceptional results from the Frank Schroeder group for the structural identification of these morphogenetic factors have been obtained in the model nematode *Caenorhabditis elegans* by 2D NMR differential approaches (DANS method) mentioned in the review by Prince and Pohnert [PRI 10]. Genetic knowledge of this model organism is used to establish a differential analysis able to reveal the essential metabolites of the lifecycle. This review, which praised the advantages of metabolomics for accelerating the discovery of mediators by bio-guided fractionation, sparked new approaches, such as those which helped elucidate the first sexual pheromone structure in diatoms [GIL 13]. This is Diproline, whose organic synthesis from Proline helped determine its absolute configuration.

In the domain of biological interactions in terrestrial plants, metabolomic studies have been used largely to detect negative or positive symbiotic regulators. Thus, in the Rhizobium-legume symbioses, the responses to Nod factors studied by a non-targeted approach revealed the importance of oxylipins as negative regulators. Metabolomic and metagenomic approaches developed equally for the study of lichen symbioses [KAM 13]. These eukaryote-prokaryote interactions are described in Chapter 3.

7.4.4. *Application of metabolomics to the study of environmental disturbances*

All environmental changes constitute a potential source, new or intensified, of directional selection on traits important for the selective value of a species. With a combination of strong selection pressures (notably in the framework of current global changes), organisms can respond in an integrative manner by adaptive responses (e.g. physiological adjustments, micro-evolutionary processes). These synergistic actions influence the state of physiological stress of organisms and populations, and eventually exert a feedback effect on the adaptive potential of species to contemporary environmental changes. Meanwhile, we hardly know the mechanisms by which the combination of environmental stress factors is translated into a physiological response, thus modifying the selective value in the natural environment. Also, distinguishing the respective parts of genetic and phenotypic responses (phenotypic plasticity) still remains difficult. However, the mechanisms of adaptation operate on the gene and organism and affect the functioning of the ecosystem. The metabolome constitutes a key level, even if little used as yet, in the study of the mechanisms of adaptation. In fact, a stress factor can induce a modification in the production of secondary metabolites, notably the volatile metabolites [LOR 10, PEN 10] which, taking into account their major role in the organism (notably chemical defense and survival), can lead to a modification of their selective value. Today, the democratization of metabolomics and new sequencing technologies applied to the genome (RADSeq) and to the transcriptome (see Chapter 6) facilitate access in a very complete manner at all organization levels, and likewise envision an integrative approach leading to a deeper understanding of acclimation and adaptation mechanisms.

7.5. Conclusions

The contribution of analytical chemistry has always been essential in chemical ecology for identifying the chemical nature of organic molecules, which are emitted often in infinitesimal quantities as signals or as defenses, but which have fundamental functions in the origin and maintenance of biodiversity and on the functioning of ecosystems. Their identification is today facilitated by technical innovations in analytical and bioorganic chemistry, helping to separate and elucidate the chemical structure of molecules sometimes at a trace level in complex matrices. These innovations have necessitated new informatics tools, which today help the manipulation of considerable sets of data, the extraction and interpretation of which are no longer possible by the sole human experimenter. This metabolomic procedure is today the key for accelerating discoveries in chemical ecology and complements the other "Omics" approaches, which also guide the discovery of these mediators in all environments.

7.6. Bibliography

[CAC 15] CACHET N., GENTA-JOUVE G., IVANISEVIC J. *et al.*, "Metabolomic profiling reveals deep chemical divergence between two morphotypes of the zoanthid *Parazoanthus axinellae*", Sci Rep., vol. 5, p. 8282, 2015.

[CUS 14] CUSIDO R.M., ONRUBIA M. SABATER J. *et al.*, "A rational approach to improving the biotechnological production of taxanes in plant cell cultures of *Taxus* spp.", *Biotechnology Advances*, vol. 32, pp. 1157–1167, 2014.

[DOR 14] DOROGHAZI J.R., ALBRIGHT J.C., GOERING A.W. *et al.*, "A roadmap for natural product discovery based on large-scale genomics and metabolomics", *Nature Chemical Biology*, vol. 10, pp. 963–968, 2014.

[FIE 02] FIEHN O., "Metabolomics – the link between genotypes and phenotypes", *Plant Molecular Biology*, vol. 48, pp. 155–171, 2002.

[FIE 07] FIEHN O., ROBERTSON D., GRIFFIN J. *et al.*, "The metabolomics standards initiative (MSI)", *Metabolomics*, vol. 3, pp. 175–178, 2007.

[GEN 12] GENTA-JOUVE G., THOMAS O.P., "Sponge chemical diversity: from biosynthetic pathways to ecological roles", *Advances in Marine Biology*, vol. 62, pp. 183–230, 2012.

[GAQ 14] GAQUEREL E., GULATI J., BALDWIN I.T., "Revealing insect herbivory-induced phenolamide metabolism: from single genes to metabolic network plasticity analysis", *Plant J.*, vol. 79, no. 4, pp. 679–692, Aug 2014.

[GIA 14] GIACOMONI F., LE CORGUILLÉ G., MONSOOR M. *et al.*, "Workflow 4 Metabolomics: a collaborative research infrastructure for computational metabolomics", *Bioinformatics*, vol. pii, p. btu813, 19 December 2014.

[GIL 13] GILLARD J., FRENKEL J., DEVOS V. *et al.*, "Metabolomics enables the structure elucidation of a diatom sex pheromone", *AngewChemlnt Ed Engl*, vol. 52, pp. 854–857, 2013.

[GOO 04] GOODACRE R., VAIDYANATHAN S., DUNN W.B. *et al.*, "Metabolomics by numbers: acquiring and understanding global metabolite data", *Trends Biotechnol.*, vol. 22, p. 245, 2004.

[GOU 12] GOULITQUER S., POTIN P., TONON T., "Mass spectrometry-based metabolomics to elucidate functions in marine organisms and ecosystems", *Marine Drugs*, vol. 10, p. 849, 2012.

[GOW 14] GOWDA H., IVANISEVIC J., JOHNSON C.H. *et al.*, "Interactive XCMS Online: simplifying advanced metabolomic data processing and subsequent statistical analyses", *Analytical Chemistry*, vol. 86, pp. 6931–6939, 2014.

[HEN 11] HENDRIKS M.W.B., VAN EEUWIJK F.A., JELLEMA R.H. *et al.*, "Data-processing strategies for metabolomics studies", *TrAC Trends in Analytical Chemistry*, vol. 10, pp. 1685–1698, 2011.

[JEN 04] JENKINS H., HARDY N., BECKMANN M. *et al.*, "A proposed framework for the description of plant metabolomics experiments and their results", *Nature Biotechnology*, vol. 22, pp. 1601–1606, 2004.

[KAM 13] KAMPA A., GAGUNASHVILI A.N., GULDER T.A. *et al.*, "Metagenomic natural product discovery in lichen provides evidence for a family of biosynthetic pathways in diverse symbioses", *Proceedings of the National Academy of Sciences*, vol. 110, no. 33, pp. E3129–E3137, 13 August 2013.

[LOR 10] LORETO F., SCHNITZLER J.P., "Abiotic stresses and induced BVOCs", *Trends in Plant Science*, vol. 15, pp. 154–166, 2010.

[NIC 12] NICOLÈ F., GUITTON Y., COURTOIS E.A. *et al.*, "MSeasy: unsupervised and untargeted GC-MS data processing", *Bioinformatics*, vol. 28, no. 17, 2012.

[NYA 13] NYAMUNDANDA G., GORMLEY I.C., FAN Y. *et al.*, "MetSizeR: selecting the optimal sample size for metabolomic studies using an analysis based approach", *BMC Bioinformatics*, vol. 14. p. 338, 2013.

[PEÑ 10] PEÑUELAS J., STAUDT M., "BVOCs and global change", *Trends in Plant Science*, vol. 15, pp. 133–144, 2010.

[POU 14] POULSON-ELLESTAD K.L., JONES C.M., ROY J. *et al.*, "Metabolomics and proteomics reveal impacts of chemically mediated competition on marine plankton", *Proceedings of the National Academy of Sciences*, vol. 111, pp. 9009–9014, 2014.

[PRI 10] PRINCE E.K., POHNERT G., "Searching for signals in the noise: metabolomics in chemical ecology", *Analytical and Bioanalytical Chemistry*, vol. 396, pp. 193–197, 2010.

[SUM 15] SUMNER L.W., LEI Z., NIKOLAU B.J. *et al.*, "Modern plant metabolomics: advanced natural product gene discoveries, improved technologies, and future prospects", *Natural Product Reports*, vol. 32, pp. 212–229, 2015.

[VIZ 14] VIZCAINO M.I., ENGEL P., TRAUTMAN E. *et al.*, "Comparative metabolomics and structural characterizations illuminate colibactin pathway-dependent small molecules", *Journal of the American Chemical Society*, vol. 136, pp. 9244–9247, 2014.

[WIN 99] WINK M., WATERMAN P.G., "Chemotaxonomy in relation to molecular phylogeny of plants", *Biochemistry of plant secondary metabolism. Sheffield Academic Press and CRC Press, Sheffield*, pp. 300–341, 1999.

[ZHA 12] ZHANG A., HUI S., PING W. *et al.*, "Modern analytical techniques in metabolomics analysis", *The Analyst*, vol. 137, p. 293, 2012.

8

Chemical, Biological and Computational Tools in Chemical Ecology

This chapter presents several techniques used in chemical ecology. In recent years, improvements in computational tools have allowed for more in-depth analyses of secondary metabolites and nucleic acids. These approaches often produce big data, making important automatic correlations with online databases. Based on performances of chemical analysis, major technical innovations have also emerged in imaging. It is, therefore, increasingly common to illustrate a given molecule with a resolution approaching the order of micrometers. This chapter presents these aspects as well as several possible axes of innovation.

8.1. Chemical tools

8.1.1. *Analytical tools of chromatography*

The characterization of secondary metabolites sent or received by different organisms constitutes a crucial, and often delicate, step in chemical ecology. Numerous advances in analytical sciences offer ecologists more or less complex alternatives.

Whatever the nature of the metabolites or the sample, the workflow is always the same and can be summarized in three main steps: sampling, analysis and structural characterization (with quantification in certain cases).

Chapter written by Nicolas BARTHÈS, Jean-Claude CAISSARD, Jérémy JUST and Xavier FERNANDEZ.

The sampling step is often crucial and a source of numerous errors. It is indeed paramount to obtain, and in a repeatable way, an extract representative of the matrix used. The latter can be of different states (gas, liquid or solid) and will be defined based on the characteristics of compounds to be analyzed.

8.1.1.1. *Sampling of volatile compounds*

The characterization of volatile compounds is often a difficult step in chemical ecology because the extraction of targeted compounds must be carried out without altering or disturbing the targeted organism(s). For many years, extraction techniques of the headspace have, therefore, supplanted techniques of solid–liquid extraction with the use of organic solvents.

These headspace techniques using stationary phases of variable selectivity are especially useful for the extraction of volatile compounds emitted by the sample [BIC 04]. Indeed, they show themselves to be increasingly simple, practical and easily automatable, as they do not require large quantities of samples and they limit the occurrence of artifacts.

Dynamic-HeadSpace (D-HS) is a highly sensitive sampling technique which is commonly used in this context. However, due to its cost (requires using a thermal desorber and a Programmed Temperature Vaporizing (PTV) injector) and complexity of implementation, the use of passive techniques have been developed, in particular, HeadSpace-Solid-Phase Micro-Extraction (SPME).

SPME is an extraction technique by adsorption of analytes on a polymeric phase deposited on a silica or stainless steel fiber (1–2 cm in length). To protect it, the fiber can retract to the interior of a needle placed on a support (Figure 8.1). The method is tried and tested and numerous polar, apolar or mixed fibers are currently available commercially. They can be reused up to 50 times, depending on their nature. By the selectivity of the stationary phase, the SPME helps concentrate the analytes of interest or eliminate interference [DUA 11].

To improve the method (sensitivity, selectivity, robustness, etc.), many techniques based on the same principle have been developed and are briefly presented in Table 8.1.

Many of these techniques are very recent and few studies have been published.

Extraction principle (reproduced with permission of Supelco Company)

Field extraction (Photo: B. Berlioz, ICN)

Figure 8.1. *Solid-phase microextraction (SPME)*

Name	Nature of the device
HSSE: HeadSpace Sorptive Extraction	Magnetic bar covered by stationary phase [BIC 05]
SPDE: Solid-Phase Dynamic Extraction	Hollow needle coated with stationary phase [KAM 07]
INDEX: Inside-needle Dynamic Extraction	Needle filled with stationary phase [BIC 04]
MonoTrap	Extraction on/in a polymeric tube/disk [SAT 09]

Table 8.1. *Example of new techniques of headspace by enrichment*

8.1.1.2. *Separation techniques of volatile compounds*

For the analysis of these compounds, Gas Chromatography (GC) is the indispensable tool. This very widespread analysis technique has been used since the middle of the last century. It spread rapidly due to its relatively moderate cost, its automation, its robustness and the large range of columns and detectors available.

More recently, it has been widely used in chemical ecology for its applications in the study of chemical communication based on volatile compounds. Numerous recent chemical ecology works have used this technique for a metabolomic approach with, for example, major results in the allelopathy domain in marine or terrestrial environments, multitrophic relations in terrestrial plants with the organisms in sustainable interactions with them, or in insect communication.

The GC technique is based on the separation of gaseous compounds according to their affinities for the stationary phase (polar columns, apolar, chiral, etc.). The analysis of liquid or solid compounds, therefore, requires heating to the gaseous state. It is the principal constraint in this technique, since we cannot use it except for the analysis of thermostable compounds. Retention time, a characteristic value of each constituent, can be used for the qualitative analysis. However, it depends on numerous operative conditions (equipment, analytical method, etc.) that make it difficult to use for the study of complex mixtures. It is generally preferable to use the retention indexes, which help to overcome this variability [KOV 58]. In fact, for a given compound, the retention index (RI) depends only on the nature of the stationary phase and not on chromatographic conditions. It is, therefore, possible to use the GC databases to attempt to identify the constituent [ADA 95]. Many of these databases are specifically dedicated to the analysis of secondary metabolites on different stationary phases.

By adapting the Mass Spectrometer (MS) column, we obtain the fragmentation spectrum of each of the eluted compounds. From the Total Ion Chromatogram (TIC), we can trace the representative chromatogram of the eluted compounds. Although this method leads to less sensitivity than conventional sensors (e.g. Flame Ionization Detector (FID)), it has become indispensable in a large number of studies.

There are many methods of ionization and separation of ions, but the electronic ionization and quadripolar analyzer have established themselves over time thanks to their robustness, price and data.

With standardized conditions, each compound presents a reproducible mass spectrum of its own. The use of spectrum libraries, therefore, helps by comparison to identify the compounds present in a mixture (see Chapter 7). Meanwhile, an unknown compound can often not be present in the available library and requires the expertise of an analytical chemist for its identification. In addition to the molecular mass of the compound, the study of the mass spectrum and the different fragments which constitute it can provide a large amount of information on its structure: this is the study of fragmentation.

Innovation requires increasingly powerful analysis techniques. In recent years, GC has progressed greatly by the development of techniques with better analytical resolution and detection limits. Thus, decreasing the length of the columns and the thickness of the stationary phases to reduce analysis time leads to the development of rapid and ultra-rapid chromatography. In fast chromatography (fast-GC), the columns are from 5 to 10 m, compared with 30–60 m for "classic" GC. The temperature programing from 15 to 50°C/min leads to a duration of analysis from 5 to 15 min compared with 45–60 min in "classic" GC for the most volatile compounds. These techniques help gain a significant amount of time but the low capacity of the columns limits the quantities of analytes which can be injected.

Sometimes, the analysis of mixtures by GC can be too complex to be entirely realized on a single stationary phase. In addition, the compounds present in the state of traces co-elute very often with more abundant analytes. To resolve this difficulty, the use of two consecutive separations on two columns of different natures helps to greatly improve the power of separation and the resolution of peaks in the selected fractions. To maximize this gain, it is, therefore, necessary to use columns with different physicochemical-based separations. The oldest bidimensional chromatography technique appeared in the 1960s and consists of transferring one or more fractions of analytes separated on the first column to the second column. This technique is called GC by heartcutting (heart-cut two-dimensional GC, GC–GC). However, this technique requires the installation, between the two columns, of a complex system of valves, counters, pressure circuits or cryogenic traps and is limited to the analysis of some critical regions of the chromatogram. The analysis time is thus extended. New developments in this domain have helped lead, in the start of the 1990s, to the development of integral bidimensional gaseous chromatography (GCxGC). One major improvement is that the totality of the sample is submitted to the separation potentials of the two chromatographic columns. It is probably one of the most promising advances in the domain of gaseous chromatography since the discovery of capillary columns.

8.1.1.3. *Detection of volatile compounds*

Classic physical detection systems (MS, FID, etc.), whether universal or specific, help obtain extensive information, but they do not furnish any data on the impact of constituents of a mixture on the target. The GC/Olfactometry

(GC/O) coupling uses a sensorial detection system (the nose for humans) as an analytical detector parallel to a physical detector and thus helps to target compounds of interest. When we position an insect antenna at the exit of a chromatograph, we speak then of electroantennography (EAG). It is the neuronal signal that indicates the molecule recognized by the insect. Meanwhile, the separation and individualized evaluation of each compound do not help to highlight the effect of the association of many compounds, sometimes in a concentration which must be well defined. The re-association of molecules in bouquet must be done later.

8.1.1.4. *Separation and detection of thermolabile compounds with little or no volatility*

Despite these performances and developments, GC does not help with the analysis of all compounds of interest. This is particularly the case for compounds which are thermally unstable or of low volatility. The derivation reactions, which help modify the structure of the analytes, are rarely used in chemical ecology because they add a step and require knowing the nature of the compounds to analyze. It is, therefore, necessary to use the techniques of Liquid Chromatography (LC).

In this technique, the extraction is generally carried out with the help of an organic solvent (or a mixture of solvents), the polarity of which is adapted to the target molecules. If the extract thus obtained is very complex, the solid-phase extraction (SPE) is particularly effective for the concentration of target analytes and/or the elimination of interferents. This technique is based on the use of varied nature of phases (silica, C8 or C18 (mainly) bonded silica, apolar copolymers, porous carbon, ion exchange resins, etc.) available in different formats (disks, cartridges, microplates). The compounds are fixed on the phase by percolation of the extract. The interferents are eliminated by solvent washing which is not sufficiently eluting to take off the target compounds. They are then released by percolation of an adapted eluting solvent.

Thin-Layer Chromatography (TLC) is a chromatographic technique that appeared in 1938 with the separation of vegetable extracts on an aluminum plate covered with silica. Since then, it has caught on quickly as a method of analysis and control. The development of High-Performance Thin-Layer Chromatography (HPTLC) marked a major advance. The use of a "high-performance" plate with stationary phases of finer granulometry helped improve all separation by increasing reproducibility. The migration systems have also evolved greatly. Glass vessels are now replaced by automatic migration systems. These instruments help control all parameters affecting the elution: rate of relative humidity, saturation and preconditioning time, migration distance and drying time of the plate. This greatly helps in increasing effectiveness and reproducibility. It is now even possible to carry out sequential developments by using a gradient mode.

Another advantage of TLC is the large number of revelators which can be used. Colored compounds are directly visible and those which are not are revealed by derivatization reaction. The results obtained with many revelators can be observed with the naked eye, which can be very useful for qualitative analyses. The development of spectrodensitometric scanners facilitates a more precise detection. It helps to detect the compounds that absorb the UV/Visible or fluorescent rays and so conduct quantitative analyses. To analyze the natural extracts, the specific revelators can be particularly useful. They do not react except with certain families of compounds.

High-Performance Liquid Chromatography (HPLC) is an analysis and quantification technique of a large number of molecules (thermolabile and/or high-molecular mass and/or polar compounds). The nature of the stationary phase and the mobile liquid phase are driven by the properties of target compounds. The separation mode is, therefore, chosen among adsorption, division, ion exchange, ion pairs or steric exclusion. The classic stationary inverse phases such as the C18 phase, more adapted to rather apolar metabolites, have been completed by Hydrophilic Interaction Liquid Chromatography (HILIC) for the more polar metabolites present in aqueous extracts.

Many types of detectors can then be used. However, contrary to the FID in GC, there is no sensitive and universal detector. The most used system is the UV/Visible detector and it requires that the compounds absorb characteristic wavelengths of an incident light. The diode array detector is more powerful. It helps scan the entire UV/Visible domain. Therefore, by spectral analysis, it is possible to carry out peak purity calculations. The Evaporating Light Scattering Detector (ELSD), and more recently the Corona detector, by its universality is increasingly used in the analysis of natural extracts. ELSD is based on the partial evaporation of the effluent as a way to obtain a mist of solid or liquid particles of the solute, which passes through a light beam. Diffused light under a determined angle is detected by a photomultiplier.

The democratization of the HPLC/MS coupling leads to a more important use of the MS detector but the conditions of non-standardized ionization do not allow using databases such as those in GC–MS.

The tandem coupling (HPLC/MS/MS) for the quantitative analyses and High-Resolution Mass Spectrometry (HPLC/HRMS) for determination of raw formulas and structural identification are increasingly used techniques.

Like GC, HPLC has progressed enormously in terms of efficiency and analysis time to lead to the U-HPLC or Ultra Performance Liquid Chromatography (UPLC). This allows, by reducing the diameters of stationary-phase particles, for larger flow ranges. It is thus possible to increase the flow, and, therefore, the speed of analysis, without altering the chromatographic separation. Today, this method has become the standard for metabolomic analyses.

Approaches combining all of the analyses are also practiced to overtake the limits of thoroughness of each individual technique. On the other hand, quality control and calibration standard samples interspersed between the analyses are essential for guaranteeing the validity of the results and limiting the drift problems among series of analyses when these are not carried out in identical conditions.

8.1.2. *Analytical approach by nuclear magnetic resonance*

Nuclear magnetic resonance (NMR) is a spectroscopic method, which consists of detecting the variations of the magnetization of certain nuclei of molecules under the action of an extremely powerful magnetic field and of an energizing electromagnetic wave. Thus, a population of atomic nuclei placed in a magnetic field is a real unique frequency oscillator. When this oscillator is subjected to the action of an electromagnetic wave, it absorbs its energy, and then dissipates it by relaxation. The relaxation makes this phenomenon observable and allows it to be studied. Thus, the NMR of observable nuclei (^{1}H, ^{13}C and ^{15}N) will allow for visualizing the return movement at the equilibrium of the atoms of the excited molecule in the magnetic field as characteristic signals. The signals thus generated and correlated, among other things, to the chemical environment of the atoms analyzed could be decrypted by the chemist who will be able to read them. This technique is mainly used to analyze and determine the structure of chemical molecules and constitutes, in this context, a technique of choice for the identification and non-selective quantification of all the metabolites in a biological system.

^{1}H NMR in solution has been used for a long time for the comparative analysis of biological samples. It has the advantages of requiring neither purification nor derivatization of the compounds to be analyzed and above all it is non-destructive. This method has especially been developed to obtain metabolic prints allowing the absolute identification and quantification of the samples' metabolites. However, it remains largely insensitive and can generate a significant number of signals, which are sometimes difficult to interpret for complex biological samples. It is also limiting for the precise chemical characterization of unknown compounds.

Two-dimensional NMR can, therefore, be an attractive alternative. Moving on to multiple dimensions allows not only better separation of information but also to establish correlations among this information. Meanwhile, this method will generate longer acquisition times and is sometimes incompatible with high-throughput analysis. Even if two-dimensional NMR facilitates the determination of numerous structures, identification is still a challenge for the minority compounds in complex samples. In this context, hyphenated methods have emerged and adapt a separative method (Liquid Chromatography) to two-dimensional NMR. This results in a pre-purification of each compound, which then will be specifically and individually analyzed in NMR. However, this approach faces other technical problems such as the elimination of the solvent used during the purification step in favor of a deuterated solvent necessary for the NMR.

In the case of tissues or cellular models, NMR High-Resolution Magic Angle Spinning (HRMAS) is used and aids the analysis of mobile metabolites in heterogeneous environments. This technique, first developed for solid NMR, uses the rotation of the sample at the magic angle (54.7°), thus helping the averaging to zero of the strong inhomogeneities present in the samples. The spectrums thus obtained on "soft" matters (gels, biological tissues, cells) samples, therefore, possess the quality of those of the high-resolution liquid spectrum. This technique helps the analysis of the crude samples, without any preparation, resulting in a significant time gain compared with other techniques.

Thus, NMR constitutes the technique of choice for metabolomics (see Chapter 7) and is, moreover, in constant evolution. Complementary to the mass spectrometry analysis, it has become an essential method for the analysis of the collection of metabolites in biological systems.

8.1.3. *Secondary metabolite imagery techniques*

In recent years, the use of measurement and physicochemical analysis instruments in imagery, the adaptation of techniques derived from the laser to biological problematics, as well as considerable progress on micrometric resolution on that equipment, has allowed unprecedented access to information on the repartition of metabolites at the cellular and subcellular levels. These techniques often cross the micrometer scale, and give access to the understanding of the mechanisms of repartition, compartmentalization and traffic of molecules of a much smaller size than macromolecules, and, therefore, sometimes to natural substances and secondary or specialized metabolites. The most used imagery in the domain of chemical ecology is the one that uses mass spectrometry. Meanwhile, other imagery techniques based, for example, on photon properties, nuclear magnetic resonance or X-rays, can prove useful because they help to image lipids, or more generally lipophile molecules which are often secondary metabolites.

8.1.3.1. *Imagery techniques based on mass spectrometry (MS)*

The main techniques of imagery based on MS use Matrix-Assisted Laser Desorption Ionization (MALDI), Matrix-Free Laser Desorption Ionization (LDI), Laser Ablation ElectroSpray Ionization (LAESI), Desorption ElectroSpray Ionization (DESI) and Secondary Ion Mass Spectrometry (SIMS) as the ionization source [BJA 14]. In all these techniques, the surface molecules of the sample are fragmented, ionized and separated depending on their *m/z* ratio. The equipment is, therefore, coupled to the mass spectrometers using, for example, a Time of Flight (TOF) detector. The profile obtained, if known previously by the experimenter, helps to identify the molecules. In MALDI, the probe is a laser beam and the resolution of the image can reach 10 µm. In DESI, the probe is a solvent and the resolution is 200 µm. Finally, SIMS attains a resolution of some tenths of micrometers with an ion beam as the probe, which renders this technique very powerful for subcellular localizations of metabolites. Unfortunately, because of the very large fragmentation of the molecules, this technique does not yet help to localize very small molecules.

Most of the time, LAESI and DESI require little or no preparation of the sample, and take place at atmospheric pressure. On the other hand, MALDI and SIMS require a very important sample preparation step as well as observation under vacuum, which prevents localizing very small metabolites. For the range of these techniques, there are already a series of works presenting localizations of metabolites, specialized metabolites and lipids with a resolution of 10–100 μm. Increasingly rapid progress will likely lead to venturing below the scale of micrometers and thus to a better understanding of intercellular traffic of metabolites as well as their secretion in the external environment.

8.1.3.2. *Imagery techniques based on photon properties*

The use of pulsed laser beam in InfraRed (IR) with confocal microscopy has led to the improvement of 2D and 3D images and to the development of microscopic techniques called nonlinear. All those improvements achieve resolution thresholds far below a micrometer and can prove useful in the subcellular localization effort of specialized metabolites [PET 10]. With this type of laser, the fluorophores are excited by the simultaneous absorption of two IR photons, each having half the energy of a UV photon, to generate fluorescence. Thus, for example, an absorption of two IR photons at 700 nm gives the same fluorescence as that of a UV photon at 350 nm, likewise an absorption of three IR photons at 900 nm corresponds to a UV phenomenon of 300 nm. The interest in the IR is that the photons penetrate further into the tissues and only meet each other at the focal point. Multi-photon fluorescence (MPF) microscopy, therefore, presents less photo-bleaching and background noise than an observation in UV rays. The observation of natural fluorophores or fluorescent markers is, therefore, better. The other interest is that this phenomenon helps to implement detection devices such as Multi-Harmonic Generation (MHG) and the Raman Effect.

When photons arrive at a molecule, they can be absorbed or scattered. In the case of MPF, the photons which arrive together are resonantly absorbed and part of the energy is transformed into fluorescence. The phenomenon is referred to as incoherent because the incident photon phase and that of the fluorescence are not correlated. To the contrary, in MHG, the incident photons are not absorbed but scattered in a single energy photon corresponding to the sum of the energies on the incident photons. The phenomenon is coherent because the phases of the excitation and emission photons are correlated. Thus, when several molecules emit in phase, harmonics are generated. Second-Harmonic Generation (SHG) and Third-Harmonic Generation (TGH) present properties which help image structures more than molecules. Thus, SHG helps, for example, to image regular but non-centrosymmetric structures such as single-phospholipid-layered membranes or crystals [CAI 12], and THG, more amorphous bodies such as lipid bodies [DÉB 06]. These techniques are, therefore, used to image crystalline metabolites, membrane exocytosis phenomena or lipid bodies susceptible to host specialized metabolites.

On the other hand, the Raman phenomenon corresponds to a loss or gain of energy. When the incident photons interact with a chemical bond of a biological molecule, they are normally scattered elastically. Meanwhile, approximately one photon for every 10^8 photons will be scattered inelastically. Its energy is that of the incident photo plus or minus the vibrational energy of the chemical bond. If the vibrational energy is added, the photon is called anti-Stokes photon; if it is subtracted, it is called Stokes photon. Potentially, each molecule, therefore, has a unique spontaneous Raman spectrum. If a wavelength of this spectrum has a very strong intensity, it is, therefore, possible to intensify the phenomenon by using two laser-beams, one of a fixed wavelength and the other variable, adjusted to the strong intensity wavelength. Imagery technologies have been developed from these phenomena. There are Coherent Anti-Stokes Raman Scattering (CARS) and Stimulated Raman Scattering (SRS) [FOL 11]. Some specialized metabolites, such as terpenoids, cyanogenic glycosides and lipids, for example, have thus been localized at the micrometer scale. The SRS technique in particular is very promising and will certainly help to image unique molecules at high resolution when the technical barriers will be crossed.

8.1.3.3. *Other imagery techniques*

In many studies, the simultaneous use of multiple techniques (CARS, THG, TOF, Raman, etc.) helps to localize secondary metabolites and/or lipophile molecules such as the triterpenoids and phytoalexins. There are also numerous other imagery techniques which could prove useful for localizing the specialized metabolites and/or lipids *in situ*. Thus, the Magnetic Resonance Imaging (MRI) technique has already shown its capacity to furnish images of lipid distribution in intact organisms with a 10 μm order of resolution.

This technique uses the detection of spins of nuclear magnetic moments to produce a virtual section in the organ. Fourier-Transform InfraRed (FTIR) imaging, X-ray imaging or even the techniques of imaging by tomography help to obtain resolutions near the micrometer. These techniques are particularly useful for identifying structural chemical elements and also, in some cases, specialized metabolites.

The improvement of resolution and the progress in the identification of molecules should increase their use in the problems of localization of specialized metabolites at the organ level as well as at the intracellular level. The interest in non-invasive, and therefore non-destructive, methods is that we can hope to follow in real time the dynamic of displacement or mobilization of specialized metabolites, and, therefore, have a new element to understand chemical communication.

Previous chemical analyses, when they generate very large quantities of data, are grouped under the term metabolomic. The genomic and the transcriptomic for their part help to establish reference sequences for the genes of an organism and to quantify the expression. They have also been hugely modernized and next-generation sequencing (NGS) can be described as revolutionary in the domain of molecular biology. The democratization of these techniques makes them also increasingly used in chemical ecology, like all the omic techniques.

8.2. Sequencing tools

The term "NGS technologies" (next-generation sequencing) encompasses all methods of high-throughput DNA sequencing developed after the classic technique of Sanger sequencing (the only one used from 1977 to 2005). These methods have considerably reduced the cost of all sequencing projects, not only by reducing the cost by nucleotide read, but also by increasing the rate of data production, and by reducing the human time necessary at each step.

By classical methods, sequencing of a eukaryotic genome would be a long-term project, requiring international funding of many millions of euros. NGS has placed this kind of project in reach of research teams, in the framework of 2–3 year funding, and allowed for the development of innovative structural and functional approaches, complementary to other omic methods for understanding how organisms and communities work in their entirety. Thanks to NGS, sequencing is no longer a goal in itself, but is a tool which integrates in a scientific approach to respond to a testable question (see Chapters 6 and 7).

8.2.1. *Principles, strengths and limitations of NGS*

8.2.1.1. *The major NGS techniques*

NGS is responsible for sequencing millions of reactions on the same solid medium, to directly measure the incorporation of each of the nucleotides at each step of the reaction and to not require DNA cloning into bacteria.

When first marketed, NGS shared the disadvantage of producing only very short individual sequences (called *reads*), usually from 35 to 50 nucleotides (nt). Currently, the achieved lengths often exceed those which were current in classic sequencing (700–900 nt), and some methods, such as Single Molecule Real-Time sequencing (SMRT), PacBio, produce reads of more than 10,000 nt (Table 8.2). Each NGS technique, developed by different companies, presents advantages, disadvantages (read length, error rates, etc.) and main applications (Table 8.2).

Technology	Read length	Error rate	Nucleotides read per run	Main applications
454	400–1000 pb	1.2%	700 Mb	*De novo* sequencing
Illumina (Solexa)	36–300 pb	0.6%	180 Gb	*De novo* sequencing, resequencing, genotyping, RNA-seq, metagenomics.
SOLiD	35–75 pb	0.06%	320 Gb	Genotyping, RNA-seq, resequencing.
Ion Torrent	200–400 pb	1.7%	10 Gb	Resequencing, RNA-seq.
PacBio	6–20 kb	14%	500 Mb	Genome finishing, metagenomics, *de novo* sequencing of bacterial genomes.

Table 8.2. *NGS technologies and their main applications*

Currently, the main remaining limitation to sequencing applications is the initial quantity of DNA required without doubt, making delicate use of the very small samples coming from microdissection or immunoprecipitation.

8.2.1.2. *Analysis of sequencing data: bioinformatics*

From the beginning, the analysis of genomic data has required the development of specific algorithms, to align the produced sequences, search sequence similarities in databases, infer phylogenetic trees, etc. However, the rapid expansion of NGS has recently challenged these methods, by their nature (short reads, with a higher error rate than the classic methods) as well as by the massive quantity of data generated. Between 2006 and 2012, it is estimated that the global volume of sequences produced doubled each year, while at the same time computing power doubled only every 18 months. If the bioinformaticians had been able to find solutions, it is important to always ask the question of data analysis, from the start of any project involving NGS: where to store the data, who will know how to process them to answer the scientific questions and where will the calculations be performed?

8.2.2. *Major domains of NGS applications*

8.2.2.1. *Genome sequencing*

The first application of classic sequencing was the establishment of reference sequences for genomes of *E. coli*, yeast, *Arabidopsis thaliana* and humans. NGS helped undertake the *de novo* genomic sequencing of a large number of new species of interest. The most widely used method is Whole-Genome Shotgun (WGS) sequencing: the extracted DNA is fragmented, each fragment is sequenced and the reads obtained are assembled by bioinformatic methods to reconstruct the genomic sequence (Figure 8.2):

– the reads are all compared, each one with the others. Those sharing a significant common sequence are fused, to build step by step a *contig*, the longest possible continuous sequence (Figure 8.2).

– using paired reads, these contigs are then linked to build *scaffolds*, sequences of which we know are of a single tenant in the genome, but which contain indeterminate regions, or *gaps* (Figure 8.2).

The current algorithms involved in NGS data use different optimizations to manipulate the massive volumes of data and to better resolve the ambiguities of the genomes. The current assemblers are MIRA, AllPaths-LG, SOAPdenovo and ABySS.

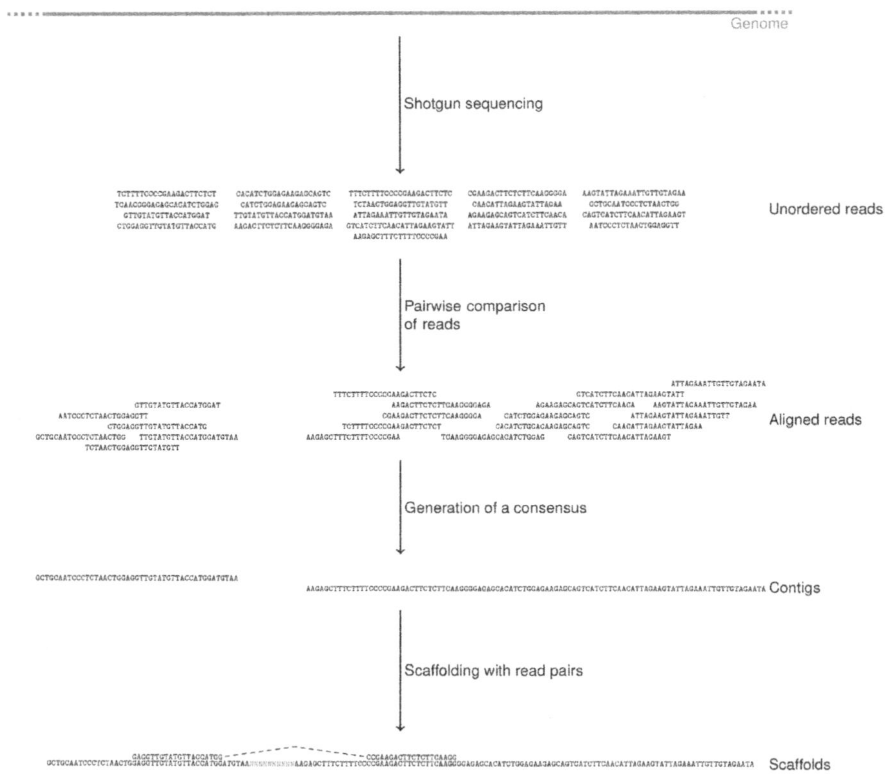

Figure 8.2. *Genomic sequence reconstituted by WGS*

Once the reference genome of a species is established, we can proceed to *resequencing* a large number of individuals, at a modest cost, to identify polymorphism: *single nucleotide polymorphisms* (SNPs), insertions–deletions (*indels*) or *copy number variations* (CNVs). These polymorphisms can be associated with phenotypes by a *genome-wide association* (GWA) study. By studying the flanking regions of the polymorphism associated with a trait, it is possible to understand the evolutionary history of the locus: hitchhiking effect or response to a selective pressure.

8.2.2.2. *Analysis of gene expression and non-coding RNAs*

After a step of reverse-transcription, it is possible to sequence the population of mRNA contained in a tissue (RNAseq). The large number of reads produced by NGS helps to quantitatively estimate the expression of each gene (Figure 8.3): the more reads we obtain corresponding to a gene, the more we consider that it is expressed. This large number of reads also allows for identifying the complete set of splicing variants of the gene. By using specific protocols to prepare the RNA library, we can also select interfering RNAs (miRNA and siRNA) and thus have access to some gene expression regulatory elements. Differences in expression patterns can also be associated with the reception of a signal or with the biosynthesis of a molecule by the tissue.

Note that methods based on hybridization (DNA, microarrays) remain valid. Their combination with NGS is cost-efficient to analyze gene expression on a large number of samples from a species without sequenced reference genome [DUB 11]. We can, for example, sequence the transcriptome from a pool of individuals using NGS, assemble it *de novo*, then design a custom microarray, and use it to analyze the samples one by one.

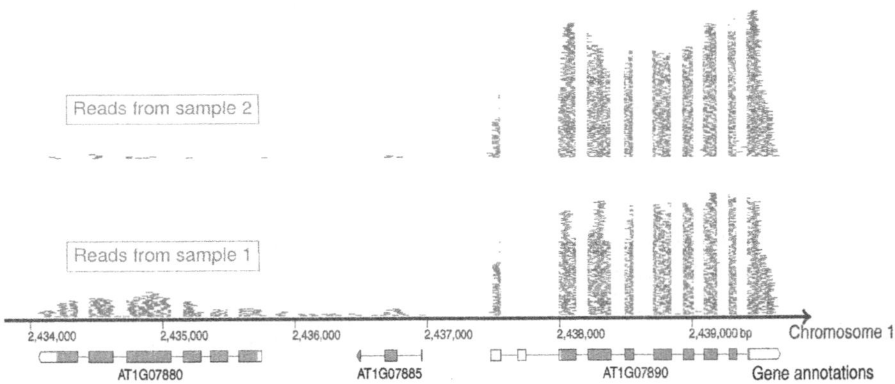

Figure 8.3. *Quantification of gene expression by messenger RNA sequencing*

8.2.2.3. *Genome live imaging*

NGS, and in particular RNA-seq, gives a static vision of the transcriptome which does not take into account the speed of transcription and the degradation of the mRNA, nor of their translation into proteins. Protocols have been developed for labeling neo-synthesized mRNA with thiouridine (4sU), before capturing and sequencing them. Native elongating transcript sequencing (NET-seq) uses the immunoprecipitation of RNA polymerase, then sequencing of linked RNA fragments from their 3' end. To study the dynamic of mRNA translation, ribo-seq selects, before sequencing, the transcript parts which are protected inside the ribosome. Finally, the ChIP-seq relies on the immunoprecipitation of a protein of interest, and the sequencing of the DNA linked to it, to identify the interaction sites, while the FISSEQ corresponds to *in situ* sequencing [LEE 14]. These techniques can prove useful to follow the dynamic of the response to a chemical stimulus, for example, and thus adapted to chemical ecology research topics.

8.2.2.4. *Community study*

Metagenomics is particularly suitable to research communities in chemical ecology, in particular those of microorganisms. Metagenomics aims at sequencing the genomic contents of an environmental sample, which can contain a large number of species. For example, DNA can be extracted from a sample of soil, seawater or intestinal contents, and then directly sequenced. The analysis of the obtained reads often aims at answering two questions: what species are present and what do they do or how do they communicate? This genomic content can be paralleled with the metabolomic content of the same sample.

The taxonomic assignation can be made by the search for targeted markers, already sequenced in numerous species, like 16S ribosomal RNA (16S-rRNA), and their comparison with databases, by a statistical analysis of the reads (e.g. tetramers frequency) or by comparison to all of the already known sequences. The interest and the difficulty of metagenomics is that a large part of the sequences obtained come from species that have never been previously observed, and are not cultivable [SHA 14].

The identification of biological functions present in the community starts by the prediction of coding sequences in the reads and their comparison to databases of known protein sequences. As a second step, the reads can be translated *in silico* into proteins on the six possible reading frames, and compared again with the databases, to find weak similarities and infer functions for the proteins coded for by the reads. These functions can be supported by the metabolomes in parallel.

The richness of metagenomic projects comes from both the characterization and the comparison of communities on already identified criteria (presence of a group of species or of biological functions specific to an environment) and of the discovery of new biological functions. Metabolomics can also give indications about selective pressures undergone by communities. It can help chemical ecology research topics by this evolutionary aspect or by identifying gene families involved in the biosynthesis of secondary metabolites.

8.3. Databases: biodiversity in silico

If the concept of biodiversity, or “diversity of the living”, is conceived more or less naturally, it is more difficult to imagine its complexity and to understand the principles which determine it, such as those at the chemistry–biology interface, for example. In fact, whatever the approaches considered, it is impossible to understand the functioning of complex biological systems if we do not order, organize and confront, in space and time, the data from the multiple undertaken studies. The question comes to define how to structure and rationalize the profusion of results acquired during *in vivo* exploration of the biodiversity with the goal of evaluating them in an integrated approach. It is this question that, in continuation of the work of naturalists and collectors of the preceding centuries, chemists and biologists have tried to answer by the conceptualization and development of databases. They do not simply to store information, by stacking results from researchers, but rationalize it to elaborate new research strategies.

Databases are a collaborative tool assuring the sustainability and traceability of the data, or knowledge associated with it, all while facilitating access and sharing in the scientific community and communities from which the studied organisms come. In fact, since 1992 with the signature at Rio of the Convention on Biological Diversity, and more recently the Nagoya protocol (in 2010), the study and evaluation of biodiversity require defining the rules of “Fair and Equitable Sharing of Benefits” access and equitable sharing of advantages between the “supplier” of resources or traditional knowledge and the “user”, whether these advantages are monetary or not. Within the framework of this arrangement, the database is one of the tools helping to establish a link between the researchers and the populations holding these resources, by registering their origin and by responding to the demands of monitoring, traceability and restitution of the acquired knowledge.

Finally, it is important to note that numerous ecosystems (coral reefs, old growth forests, mangroves, maquis/bush, etc.) disappear each year, or are extremely weakened by human activities, which induces an inexorable impoverishment of biodiversity. The work of scientific inventory, notably by the implementation of databases, as reference tools for measuring this impact, is by consequence essential because it will compile the "photos" of our natural heritage in time.

The databases currently used in chemical ecology have been listed in a non-thorough manner in this part.

8.3.1. *Databases of chemical compounds and general ecology*

– The database of *Global Biodiversity Information Facility* (http://www.gbif.org/):

This international database created at the initiative of the OCDE collects taxonomy, ecology and molecular raw datasets to give visibility to the inventories of biodiversity. In August 2014, this base listed 500 million raw data dated and geo-referenced.

– The database of the *World Register of Marine Species* (http://www.marinespecies.org/):

This international database was initiated by the European Register of Marine Species and references the names of marine species known to date. Databases maintained by other organizations such as the Flanders Marine Institute regularly come to enrich the WoRMS database to maintain consistency of the taxonomic nomenclature.

– The *Pherobase* (http://www.pherobase.com):

This collaborative database references emitted chemical compounds, pheromones, but more generally all chemical mediators, by species. The data mainly references compounds emitted by plants and insects for historical reasons but the base is regularly updated with data from vertebrates and marine organisms based on voluntary contributions.

– The *MarinLit* database (http://pubs.rsc.org/marinlit):

This paid database created in the 1970s by professors John Blunt and Murray Munro of the University of Canterbury, New Zealand, is dedicated to natural marine products. This compiles the collection of published results in the literature and references taxonomic, ecological and also chemical classification and biological activity. Although paid, it is a valuable tool for all ecologist or chemist researchers working in the natural substances of the marine domain.

– The *Metlin* metabolics database (https://metlin.scripps.edu/index.php):

This collaborative database contains information on numerous usual metabolites. Including mass spectrometry data on metabolites, it is the default database for automatic searches of the LC–MS statistical analysis tool "Xcms online".

– The *Galaxy* platform (http://galaxyproject.org/):

This platform offers all the functionality necessary to work on the databases. It integrates all statistical tools and all querying tools of banks of data in the same suite of analyses.

8.3.2. *Databases for the omics that can be used in chemical ecology*

The National Center for Biotechnology Information (*NCBI* – http://www.ncbi.nlm.nih.gov/).

The European Bioinformatics Institute (*EBI* – http://www.ebi.ac.uk/).

The Ensembl project: genome databases for vertebrates and other eukaryotic species available freely online (http://www.ensembl.org/index.html &http://metazoa.ensembl.org/index.html).

The Kyoto Encyclopedia of Genes and Genomes (*KEGG* – www.genome.jp/kegg/).

The Plant Omics Data Center (*PODC* – http://bioinf.mind.meiji.ac.jp/podc/).

The *EMBL-EBI* provides freely available data from life science experiments covering the full spectrum of molecular biology (https://www.ebi.ac.uk/metagenomics/).

Other databases linked to chemical ecology and the metabolomic are provided in Table 8.3.

Mass spectral databases	
HMDB (Human Metabolome Database)	www.hmdb.ca/
Scripps Metlin	http://metlin.scripps.edu/
KNApSAcK	http://kanaya.naist.jp/KNApSAcK/
MassBank	www.massbank.jp/
GMD (Golm Metabolome Database)	http://gmd.mpimp-golm.mpg.de/
FiehnLib (Fiehn Metabolome library)	http://fiehnlab.ucdavis.edu/projects/FiehnLib/index_html
NIST	www.nist.gov/index.html
MMCD	http://mmcd.nmrfam.wisc.edu/
Metabolite databases	
SWMD (Seaweed Metabolite Database)	www.swmd.co.in/
ChEBI	www.ebi.ac.uk/chebi/
DrugBank	www.drugbank.ca/
PubChem	http://pubchem.ncbi.nlm.nih.gov/
MarinLit	www.chem.canterbury.ac.nz/marinlit/marinlit.shtml
LIPID MAPS	www.lipidmaps.org/
Chemspider	www.chemspider.com/
KEGG	www.genome.jp/kegg/
MMCD	http://mmcd.nmrfam.wisc.edu/
(Meta)Genomic databases	
Megx	www.megx.net/
Camera	http://camera.calit2.net/
JGI	http://genome.jgi-psf.org/
GOLD	www.genomesonline.org/cgi-bin/GOLD/index.cgi
Data processing links	
XCMS	http://metlin.scripps.edu/xcms/
MetAlign	www.pri.wur.nl/UK/products/MetAlign/
Galaxy Workflow4Metabo	http://galaxy.workflow4metabolomics.org/
MZmine2	http://mzmine.sourceforge.net/
MetaboAnalyst	www.metaboanalyst.ca/MetaboAnalyst/

metaP-Server	http://metabolomics.helmholtz-muenchen.de/metap2/
MetAtt	http://metatt.metabolomics.ca/MetATT
Metabolome Express	www.metabolome-express.org
AMDIS	www.amdis.net/
SpectConnect	http://spectconnect.mit.edu/
SpectConnect	http://spectconnect.mit.edu/

Table 8.3. *Metabolomic database links*

8.4. Conclusions

The dramatic progress of "omics" techniques and their democratization has led researchers to integrate them in all biological disciplines and particularly in chemical ecology. These techniques now generate data so numerous and massive that the problem becomes one of the *big data*: where to store them and how to process them? This problem is crucial because its response will probably depend on the advancement of new concepts of chemical communication among living organisms. This will certainly also help to compare dynamics of genetic, or even cellular, responses and, therefore, to better understand this language of nature, which is, without any doubt, the oldest of all languages.

8.5. Bibliography

[ADA 95] ADAMS R.P., "Identification of essential oil components by gas chromatography/mass spectroscopy", *Society for Mass Spectrometry*, vol. 16, 1995.

[BIC 04] BICCHI C., CORDERO C., RUBIOLO P., "A survey on high-concentration-capability headspace sampling techniques in the analysis of flavors and fragrances", *Journal of Chromatographic Science*, vol. 42, pp. 402–409, 2004.

[BIC 05] BICCHI C., CORDERO C., LIBERTO E. *et al.*, "Impact of phase ration, polydimethylsiloxane volume and size, and sampling temperature and time on headspace sportive extraction recovery of some volatile compounds in the essential oil field", *Journal of Chromatography A*, vol. 1071, pp. 111–118, 2005.

[BJA 14] BJARNHOLT N., LI B., D'ALVISE J. *et al.*, "Mass spectrometry imaging of plant metabolites – principles and possibilities", *Natural Product Reports*, vol. 31, pp. 818–837, 2014.

[BOR 13] BORISJUK L., ROLLETSCHEK H., NEUBERGER T., "Nuclear magnetic resonance imaging of lipid in living plants", *Progress in Lipid Research*, vol. 52, pp. 465–487, 2013.

[CAI 12] CAISSARD J.-C., OLIVIER T., DELBECQUE C. *et al.*, "Extracellular localization of the diterpenesclareol in clary sage (*Salvia sclarea* L., Lamiaceae)", *PLoS ONE*, vol. 7, p. e48253, 2012.

[DÉB 06] DÉBARRE D., SUPATTO W., PENA A.-M. *et al.*, "Imaging lipid bodies in cells and tissues using third-harmonic generation microscopy", *Nature Methods*, vol. 3, pp. 47–53, 2006.

[DUA 11] DUAN C., SHEN Z., WU D. *et al.*, "Recent development in solid-phase microextraction for on-site sampling and sample preparation", *Trends in Analytical Chemistry*, vol. 30, pp. 1568–1574, 2011.

[DUB 11] DUBOIS A., REMAY A., RAYMOND O. *et al.*, "Genomic approach to study floral development genes in *Rosa sp.*", *PLoS ONE*, vol. 6, p. e28455, 2011.

[FOL 11] FOLICK A., MIN W., WANG M.C., "Label-free imaging of lipid dynamics using coherent anti-Stokes Raman scattering (CARS) and stimulated Raman scattering (SRS) microscopy", *Current Opinion in Genetics and Development*, vol. 21, pp. 585–590, 2011.

[KAM 07] KAMPHOFF M., THIELE T., KUNZ B., "Influence of different extraction parameter on a solid-phase dynamic extraction for the gas chromatographic determination of d-limonene degradation products by using a fractional factorial design", *Journal of AOAC International*, vol. 90, pp. 1623–2627, 2007.

[KOV 58] KOVATS E., "Gas chromatographic characterization of organic compounds. I. Retention Indexes of aliphatic halides, alcohols, aldehydes, and ketones", *Helvetica Chimica Acta*, vol. 41, pp. 1915–1932, 1958.

[LEE 14] LEE J.H., DAUGHARTHY E.R., SCHEIMAN J. *et al.*, "Highly multiplexed subcellular RNA sequencing in situ", *Science*, vol. 343, pp. 1360–1363, 2014.

[PET 10] PETITBOIS C., "Imaging methods for elemental, chemical, molecular, and morphological analyses of single cells", *Analytical and Bioanalytical Chemistry*, vol. 397, pp. 2051–2065, 2010.

[SAT 09] SATO A., SOTOMARU K., TAKEDA M. *et al.*, "A novel approach for sample adsorption and extraction using a monolithic hybrid adsorbent as a new generation medium", *LC–GC Europe*, pp. 31–32, 2009.

[SHA 14] SHARPTON T.J., "An introduction to the analysis of shotgun metagenomic data", *Frontiers in Plant Science*, vol. 5, p. e209, 2014.

9

Academic and Economic Values of Understanding Chemical Communication

Chemical mediators are small molecules of natural origin, which are involved in the interactions between living organisms and their environment. They can be at the focus of investigations in biology and biochemistry because they have evolved under pressure of natural selection for regulating cellular processes or interacting with specific receptors.

The progress realized in chemical ecology during recent decades has led to awareness growing from the interest, from an economic point of view, to better decrypt chemical communication. In this chapter, we put forward concrete applications of directed studies in chemical ecology, practical applications (pest management, development of new biopesticides, cosmetics or medicine, control of marine organisms responsible for fouling, phytoremediation, etc.), as well as fundamental applications (new biosynthesis pathways, biomimetic synthesis, identification of new biological targets, etc.).

9.1. Nature as a model

Since life began on the planet, around 3.8 billion years ago, it has had time to perfect forms and to optimize them over millions of years. On a scale of time measured in billions of years, life is built on adaptive and sustainable strategies. Man has also evolved in his environment by observing and being inspired by nature without so far rationalizing his reflection. Current

---

Chapter written by Bernard BANAIGS, Ali AL MOURABIT, Guillaume CLAVÉ and Claude GRISON.

researchers try to understand natural solutions developed by living organisms at the most fundamental level, to dissect the functioning of systems capable of producing a remarkable phenomenon and copying it. This is the principle of biomimetics or bio-inspiration [BEN 98]. The airfoil of a plane/wings of a bat, self-cleaning glass/lily pad, ship hulls or water-repellent fabric/shark skin, Velcro/burdock fruits, technological components in bees' nests/hexagonal cells of the hive, passive cooling in buildings/termite mound cooling, super adhesive without glue/gecko feet, coral/dental prosthetics, nanostructured material/diatoms, sonar and radars/acoustic systems of cetaceans and bats, and wind turbines/whale fins are all examples of technological solutions inspired from the genius of nature!

These discoveries inspired from nature are mostly from the material sciences. Examples of bio-inspiration at the molecular level are rarer; antifreeze proteins/fish of the Southern Ocean, enzymes/hyperthermophilic bacteria, and surgical glues/byssus of mussels are some of them.

This procedure which consists of copying and mimicking natural processes begins to be integrated in the studies carried out in chemical ecology, an integrative science in itself. The applications which result directly from work developed in chemical ecology are increasingly frequent, but still remain too anecdotal and restricted to easily observable phenomena. Meanwhile, the sources of inspiration are infinite. Nature is a formidable reservoir of ideas.

In chemical ecology, the general procedure consists schematically of dissecting a dual or multitrophic interaction, mediated by chemical messengers in a given ecosystem, and after having identified the emitter and the receptor, decoding the chemical language used in the transfer of information; this is to try to understand the mode and tools of production of chemical mediators, and to identify the biochemical receptors involved in reception of messages, which we will come to see in the following chapters.

Teams of researchers, ecologists, biologists and chemists involved in this fundamental procedure of observation and understanding of natural procedures at the molecular scale, imagine increasing numbers of new fields of application by integrating new questions:

– Does the chemical mediator characterized, selected from thousands of years of evolution and co-evolution for its effectiveness and its specific activity on the target organism, present an interest for man (medicine, biopesticide, antifouling, etc.)?

– Is its mode of action original? Can it lead to the discovery of new potential cellular receptors, or targets for new therapeutic agents?

– How to obtain this mediator, sometimes with a very complex structure, in a sufficient quantity to confirm its role in the environment or to perform pharmacological tests (new synthesis methodologies, biomimetic synthesis, etc.)?

– By which metabolic pathway and by what partner (host, symbiont, parasite) was the mediator produced? Can it be obtained in quantity through biotechnology?

– And more generally how the fundamental understanding is acquired upstream, can it supply the ecotechnologies, and respond to societal demand and meet the challenges of sustainable development?

These approaches will help chemists use bio-inspired procedures and develop more effective chemistry, and the ecologist to imagine solutions which are able to favor the circular economy while following the criteria of sustainable development.

9.2. Nature as a model for development of new molecules of interest

For thousands of years, man has known how to find in nature, often in the plant kingdom, the means to care for himself. In the western world, doctors and pharmacists have not stopped looking to identify in an active extract, which can be a very complex mixture, the molecule(s) responsible for the activity. But, most often, this quest is carried out by screening at a large scale plants, microorganisms or marine organisms. This type of approach, which has borne fruit, requires substantial resources that only the pharmaceutical or phytopharmaceutical industry, or large research centers can mobilize, the structure–function link being established *a posteriori*.

Another complementary approach is to look at functional metabolites, molecules that have been clearly shown to play an ecological role, a role of mediator in intra- or interspecific relations, and which were susceptible to bind in a specific manner to a cellular receptor. The study of the mechanism of action of this molecule and the search for its receptor could then help orient tests to the judiciously blindly, rather than chosen targets of interest.

If the molecule proves to be an effective ligand of interest, it will be synthesized in the laboratory. The intellectual process to draw on the biosynthetic pathways is increasingly involved in the multistep synthesis of complex molecules. The production, at a laboratory but also an industrial scale, through biomimetic pathway helps to partially replace petrochemical procedures, based on the use of fossil resources, by unique, inexpensive and often more selective procedures (synthesis of a given isomer or stereoisomer).

If the receptor of the mediator is original (new protein), or has not been described to induce the cellular response observed in nature, it can then constitute a new target for the research of new ligands, natural or synthetic.

We offer in this chapter several examples.

9.2.1. *From chemical mediators to new bioactive structural archetypes*

Microorganisms, small eukaryotes (fungi, algae, etc.) and prokaryotes, have developed ingenious and often complex strategies to expand their capacity to develop in an ecosystem and persist to the detriment of their diverse competitors (see Chapter 3). The production of chemical weapons, such as the *modified peptides* (NRPs and RiPPs) is one of the most effective strategies. It is triggered when the conditions of life become unfavorable.

The recent expression "ecological therapeutic lead discovery" encompasses the changing attitudes in this field particularly applicable to the study of the molecular dialog between microorganisms and can lead to the discovery of new antibiotics [SAN 13].

9.2.1.1. *The lasso peptides*

The lasso peptides [LI 14] are bacterial defense peptides, which present a lasso structure, that is, in the form of a loop similar to the lasso of a cowboy. This structure is achieved by the formation of a macrocycle (solid arrow in Figure 9.1) at one end of the amino acids chain, in which is inserted and irreversibly trapped the other end (dashed arrow, Figure 9.1). In nature, these peptides are synthesized in different bacterial genera from a linear precursor by the action of two maturation proteins. Researchers are notably interested in two molecular mechanisms governing the biosynthesis of a lasso peptide called microcine J25 or MccJ25 [EBR 13], synthesized by the *Escherichia coli* bacteria, and inhibitor of bacterial RNA polymerase. The objective is to see if the biosynthetic machinery of MccJ25 can be used to provide other lasso peptides and cyclic branched peptides, thus opening perspectives into the engineering of bioactive peptides from the lasso structure [PIS 15]. Particularly stable (acidity, temperature and enzymatic degradation) lasso peptides are in fact good candidates for applications in different domains (food, human or animal health), where the demand is becoming increasingly urgent.

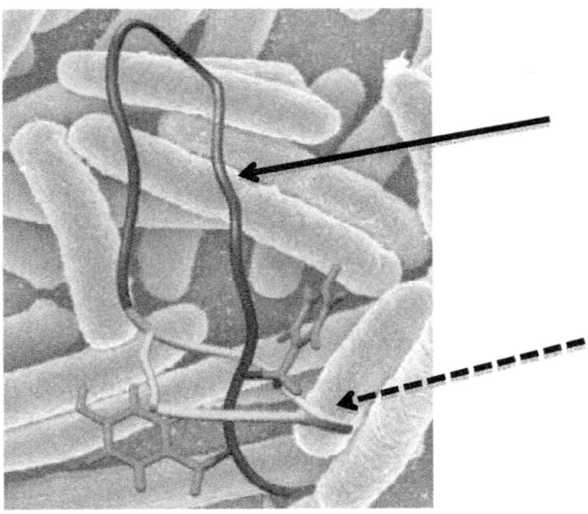

Figure 9.1. *Microcine J25, a lasso peptide produced by the enterobacteria Escherichia coli (solid arrow: macrolactame cycle, dashed arrow: peptidic tail) (http://www2.cnrs.fr/presse/communique/2651.htm?debut=48)*

9.2.1.2. *Peptaibols*

Peptaibols are small, linear neurotoxic peptides, of molecular weight between 500 and 2200 Da. They are present in diverse species of filamentous fungi of terrestrial or marine origin, principally of the genus *Trichoderma*. They are characterized by the presence of original α,α-dialkylated amino acids such as α-aminoisobutyric acid (Aib) and isovaline (Iva) and by a C-terminal amino alcohol residue. Peptaibols are biosynthesized through a non-ribosomal pathway (NRPS). The essentially helical conformation of peptaibols, with a marked amphipathic character, gives them insertion properties in cellular membranes, leading to the formation of structures such as pores or transmembrane voltage-dependent ion channels. In the early 2000s, it was shown that alamethicine, a 20-amino-acid peptaibol, was an elicitor of biosynthetic jasmonates and salicylate pathways. Peptaibols act as signal molecules to stimulate the natural defense mechanisms of plants. These elicitors can serve as a model for the development of new processes of replacement of synthetic pesticides used by farmers to control pathogens [BES 06]. The entry criteria into the sustainable agriculture market, including case of synthesis, activity at very low doses, non-toxicity and biodegradability are found together in these molecules. New strategies of synthesis of recently developed peptaibols [BEN 14] will allow easier access to this type of molecule.

9.2.1.3. *Lipopeptides*

Numerous naturally occurring lipophilic peptides (for review, see [BAN 14]) are commercialized as antibiotic drugs (actinomycin, vancomycin, tyrocidine, gramicidin, daptomycin, etc.), antifungal drugs (caspofungin), cytostatic drugs (bleomycin, epothilone) or immunosuppressor agents (cyclosporine A). The ecological role of these molecules, isolated mostly from microorganisms, is not yet completely elucidated, but an increasing number of works show that they are expressed in response to the presence of competitors. The great diversity encountered in these peptides is not only due to the enormous number of possible combinations from the 20 essential amino acids, but also due to the frequent presence of non-proteinogenic amino acids resulting from post-translational modifications (ribosomal peptide synthesis and post-ribosomal modifications), from non-ribosomal synthesis (NRPS) and from a combined process in which the NRPS and polyketide synthases (PKS) are involved. Many dozens of lipopeptides are currently in clinical trials for antibiotic or antitumoral activities. Lipopeptides have applications in other domains besides human health. Thus, antibiotic lipopeptides, produced by an isolate of *Bacillus subtilis* obtained from the xylem of an American elm, inhibit many fungal phytopathogens, including the fungi *Ceratocystis ulmi* which is transported by scolytes and responsible for Dutch elm disease, which has already destroyed numerous populations of elms. It has also been shown that surfactin-type lipopeptides played an auto-inductor role in quorum sensing, by activating a pathway of cell differentiation, bacterial cannibalism, sporulation or the formation of biofilm. Chemical diversity and physicochemical (surfactants, emulsifiers, etc.) and biological properties of the lipopeptides were at the origin of the creation of a biotechnology start-up, Lipofabrik, whose main activity is the production, formulation and commercialization of lipopeptides (iturins, surfactin, fengycin, etc.) from *B. subtilis*.

9.2.1.4. *Quorum sensing and quorum quenching molecules*

Quorum sensing is one of the cellular communication systems, which allows modulation and synchronization of gene expression in bacterial populations. The autoinducer is a molecule secreted by bacteria, which diffuses freely in the environment and can pass through the wall and bacterial membrane. When the bacterial population density rises, the autoinducer concentration increases until the formation of a complex autoinducer/transcription factor in the bacteria. The formed complex activates a gene, inducing a response: luminescence, virulence, formation of biofilms, etc. Quorum sensing concerns numerous domains such as medicine, agriculture, aquaculture, water treatment or remediation. Quorum sensing is, for example, an interesting target for future antibiotics which no longer kill bacteria but prevent them from causing damage by disorganizing them. The formation of biofilms also constitutes a medical (organs affected by diseases, implants) as well as environmental (pollution) challenge, but the mechanisms, while universal, are not always well known, especially in the natural environment. In interspecific competition, between bacteria/bacteria or bacteria/eukaryotes, certain species produce molecules inhibiting the autoinducer (quorum quenching) action; much current research focuses on the characterization of these molecules of interest [KAL 15].

9.2.1.5. *Mycosporine-like amino acids (MAAs)*

UV constitutes, from an energy point of view, the most active part of solar radiation to which living organisms are subjected. They are, therefore, responsible for the vast majority of the deleterious effects linked to solar exposure. UV in fact provokes irreversible lesions on DNA. Powerful UV radiation induces in numerous living organisms, terrestrial or marine, the production of small molecules, mycosporine-like amino acids (MAAs), playing the role of chemical photoprotective agents. These natural UV-absorbents are particularly important for numerous microorganisms, including algae and sessile invertebrates of coral reefs that live in transparent waters at a shallow depth, and are therefore, exposed to intense UV radiation during long hours. More than 20 MAAs have been described and serve as a model of synthesis of new chemical filters for applications in cosmetics (sun screen). Studies have also been conducted for evaluating the biomass constituted by MAA-rich invasive red algae, [BED 14].

9.2.2. *Biosynthesis and biomimetic synthesis*

Beyond unimaginable chemical structures and unsuspected biological activities of natural substances themselves, the biosynthetic pathways at the origin of their formation are also a source of inspiration. This could promote their synthesis in a biomimetic fashion, which could be more effective, faster and less energy consuming [POU 11]. The synthesis of molecules of biological interest eliminates the need to use natural sources. Concerning the production of chemical substances, organic chemists are more and more concerned with using effective reactions and catalysts to achieve complex transformations. Academic and industrial laboratories have realized that nature constitutes a significant source of inspiration for conceiving and elaborating molecules and making them available to biologists even when these are of an unequaled complexity. A certain number of examples testify to the validity of chemical solutions contributed by biomimetic reasoning.

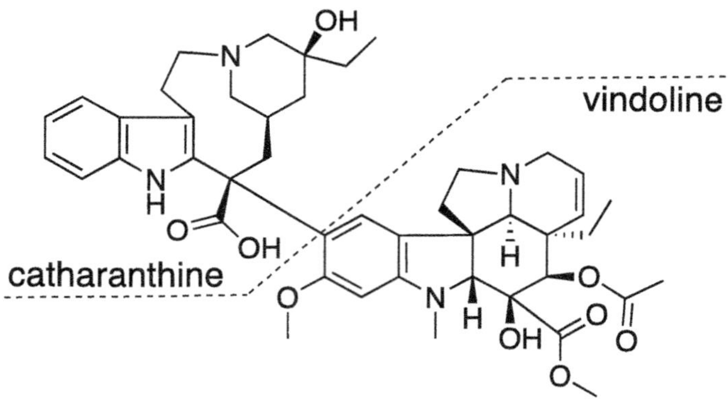

Figure 9.2. *Vinblastine, an anti-mitotic alkaloid isolated from the* Halichondria okadai *sponge*

Based on a biogenetic hypothesis, the dimerization of vindoline and catharanthine into vinblastine (extracted from Madagascar periwinkle, *Catharanthus roseus*) [ROB 98], a synthetic scheme was developed using a crucial biomimetic coupling for the development of the molecule as a drug (see Figure 9.2). Biomimetic coupling simplified synthesis and allowed the direction of efforts toward the synthesis of new analogs for potential antimitotic activities.

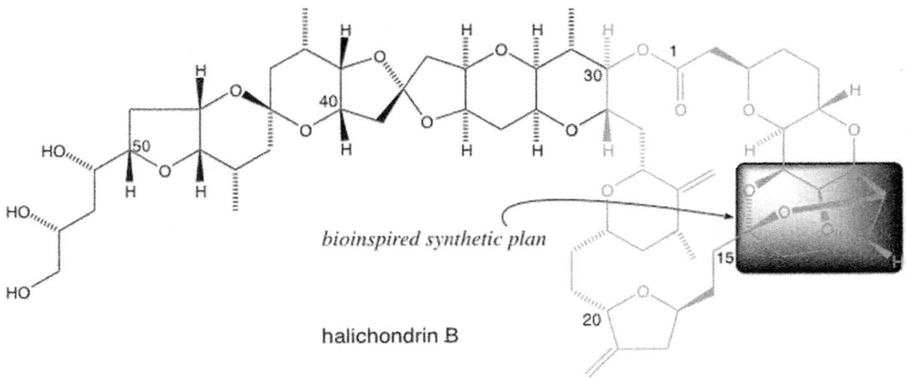

Figure 9.3. *Halichondrin B, a source of bioinspiration for the chemist (structure, biological activity and reactivity)*

Halichondrin are a series of highly cytotoxic cyclic polyesters isolated from the *Halichondria okadai* sponge (e.g. halichondrin B, Figure 9.3). These are "spindle poison"-type anti-tumoral agents; they perturb the mitosis by inhibition of polymerization of tubulin in microtubules, by binding to a tubulin dimer with a strong affinity. Structural diversity is brought about by variations in the length of the carbon chain (C50-C54 part) and the degree of oxidation of the bridged tricyclic system (C10-C15 part). This acetal polycyclic site, essential for biological activity, was particularly difficult to obtain by synthesis. The total synthesis of halichondrin constitutes a real challenge for the organic chemist. At first sight, the product does not seem easy to handle to bring it into clinic. Many teams have started in the exercise of its synthesis in an initially academic framework. The Kishi group at Harvard University, in collaboration with the National Cancer Institute (NCI) and Eisai, a Japanese pharmaceutical company, achieved total synthesis of halichondrin and the study of structure–activity relations of the synthetic intermediates. The evaluation of antitumoral activities of the collection of more or less complex and modified intermediates culminated with the discovery of one of the intermediates, which became a medicine marketed in November 2010, Eribulin mesylate (E7389, Halaven™) [YU 11]. Before the synthesis, the program began with the ambitious idea of extracting the natural resource using a massive collection (200 kg of sponge) to then extract from this 300 mg of active halichondrin. The synthesis program initiated by Kishi was then able to reach it full potential, and the battle to optimize each of the steps of synthesis became crucial. The construction of the polycyclic heart of halichondrin and its truncated analog is one of the important steps among the 35 steps necessary for the synthesis of Eribulin, active at the nanomolar range of concentration. Bioinspired reasoning was a great help in assembling the polycyclic cage. The formulated biosynthetic hypotheses for halichondrin B have shown the probable linear intermediate (see Figure 9.4). In fact, this linear intermediary corresponds to a standard polyketide assembly, coded by known and identified gene modules.

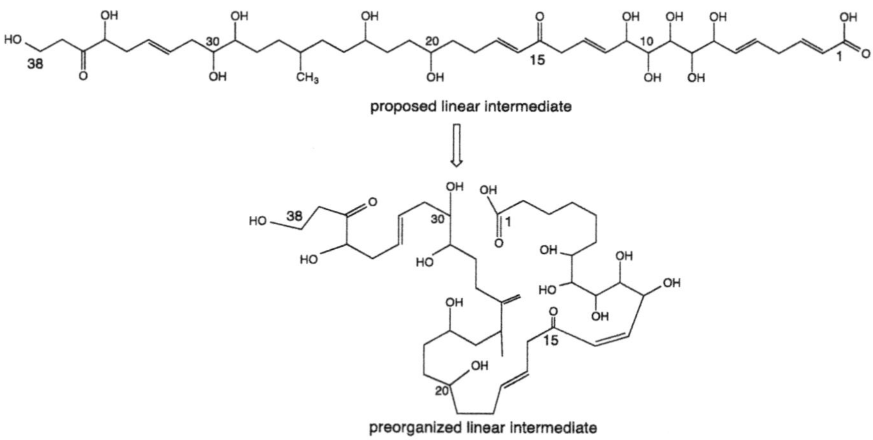

Figure 9.4. *Linear intermediary represented as extended and pre-organized before cyclization*

This skeleton transforms into Eribulin after preorganization in adequate conditions (see Figure 9.5). These are the collection of biosynthetic considerations which have resulted in a synthetic scheme including the key biomimetic steps. Eribulin is an anticancer drug used in the treatment of metastatic mammary carcinoma.

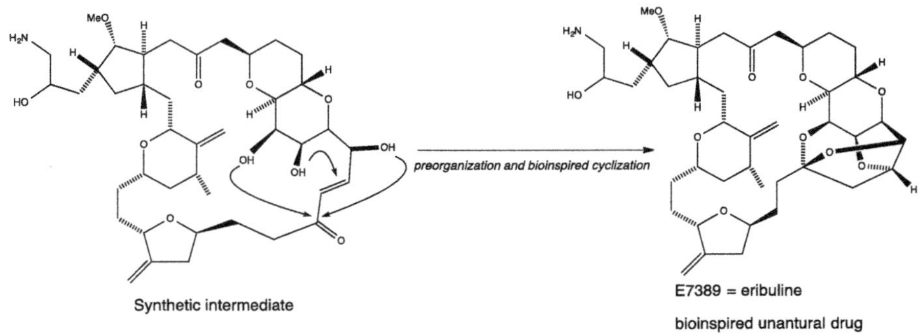

Figure 9.5. *Cyclizations of the bridged tricyclic system (C10–C15 part)*

9.2.3. *Chemical mediators and ligand/receptor interactions: to the discovery of new cellular receptors and biochemical tools*

Parallel with the discovery of new bioactive structural archetypes from 1960 to 1980, following large-scale screening, the notion of ligand and receptor was affirmed. The active compound, the ligand, induced a biological response by linking to the receptor, a membrane or cellular protein, in a specific manner. The isolated receptor, if it is of interest for the pharmaceutical or phytopharmaceutical industry, will then serve as a macromolecular target for the research of new exogenous ligands, natural or synthetic. Biological diversity generates an extraordinary chemical diversity, and it is by studying the mode of action of certain compounds having a novel structure that new receptors, targets for certain pathologies, have been discovered [HON 11].

Historically, studies conducted in ecotoxicology on the mode of action of tetrodotoxin and saxitoxin, two paralytic marine phycotoxins, have played a major role in the development of the sodium channel concept, and more generally, the transmembrane channels. Owing to their strong selectivity and affinity, these two molecules were used as molecular probes to isolate, purify and classify the sodium channels. They are both commercialized, as a dozen other toxins (brevetoxin/voltage-dependent Na^+ channels, curacin A/tubulin, okadaic acid/protein phosphatase, staurosporine/protein kinase, etc.), by various companies as molecular probes or biological tools. All these molecules are used in cellular biology, biochemistry or pharmacology, to explore in detail certain cellular mechanisms or to define the function of a protein.

Fumagillin is an antibiotic agent produced by a fungal strain and used in human clinical medicine in the treatment of infections in immuno-depressed patients, and in apiculture. It was also shown that this terpenoid possessed an angiogenic activity by inhibition of a methionine amino peptidase (MetAP) [SIN 97]. MetAP has become a new target in research into new antitumoral compounds.

Finally, Diazonamide A is a complex pseudopeptide, isolated from an ascidian and possessing antimitotic activity, as well as taxol or vinblastine. But contrary to these two antitumoral drugs, diazonamide does not bind to tubulin. Searching for its cell receptor helped to highlight a completely original mode of action, the implication of a mitochondrial enzyme, the ornithine δ-amino transferase (OAT), in mitosis, and to propose this protein as a new target for antitumoral agents' development. OAT was not known to play a role in the regulation of the mitotic spindle [WAN 07].

9.3. Chemical ecology and sustainable development

Most living organisms permanently intercept indispensable information to find a suitable habitat to develop, to find their prey or their partner, to protect themselves or avoid predation. This information is often transported by chemical mediators (see Chapter 1). These mediators therefore play a central role in structuring and functioning of an ecosystem, be it wild or domesticated by man. Any perturbation (acidification, global warming, pollution, etc.) will lead to a degradation of biotic relations, which can be due to:

– the weakening of the emitter organism (lower emissions of chemical mediators) and/or the receptor organism (e.g. presence of endocrinal perturbators), leading to a bad perception of the signal eventually with lower phenomena of attraction (pollination, mating, larval recruitment, etc.);

– the interference of an emitted signal (e.g. presence of info-disruptors); or

– an inhibition of the ligand/receptor interaction (e.g. saturation of receptors by xenobiotic antagonists).

All these deleterious effects can of course accumulate and eventually lead to an anthropized system.

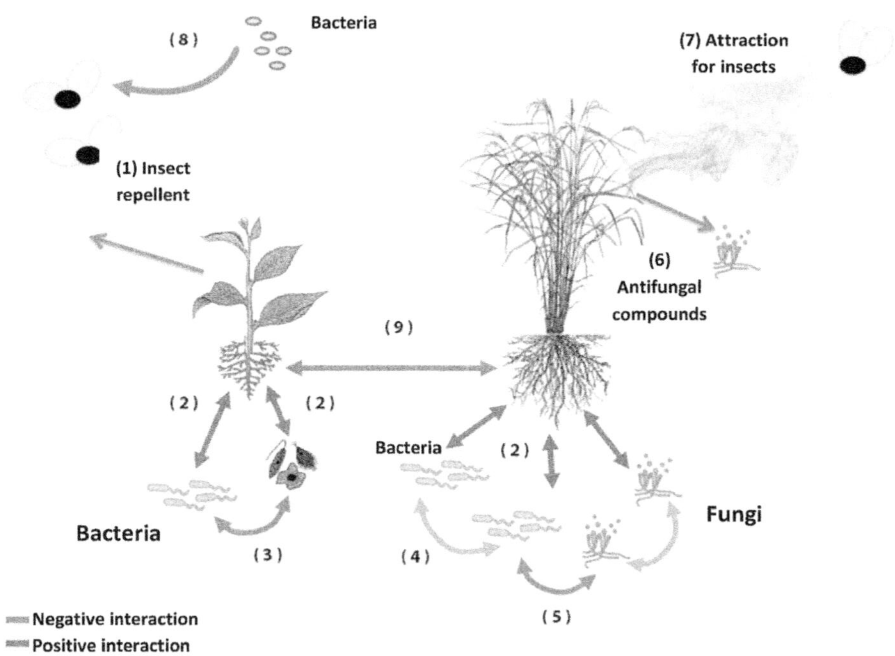

Figure 9.6. *Examples of plant/insect/microorganism biotic interactions (after [BER 15]); (1) compounds emitted by the plant with repulsive effect for pest insects; (2) compounds emitted by the plant with attractive effect for symbiotic microorganisms or phytopathogens; (3 and 5) compounds emitted by tolerated microorganisms with a negative effect for other microorganisms; (4) quorum sensing bacterial compounds; (6) compounds emitted by the plant with a negative effect for microorganism phytopathogens; (7) compounds emitted by plants to attract pollinators; (8) compounds emitted by entomopathogen bacteria; (9) phytotoxic compounds emitted by the plant to limit the growth of competing plants*

By exploring plant/plant, plant/pest and plant/auxiliary biotic interactions (see Figure 9.6), to stay only on the continental environment, and by dissecting the language of nature, researchers can mobilize the acquired understanding to conceive sustainable protection strategies against insect pests, pathogens, disease carriers and biological invasions or to imagine solutions for the rehabilitation of strongly anthropized sites. These ecotechnological approaches lead quite naturally to the creation of start-ups backed by a public research organization. New culture strategies such as push–pull, devices that repel pests of the culture and which attract them to the plant traps or which attract the predators of these pests, tend to develop.

9.3.1. *Bio-control*

Intensive agriculture is characterized by the extensive use of pesticides which, according to all available studies, seriously affect our environment and can have grave consequences for our health. The development of natural products capable of effectively replacing them constitutes a challenge. Sustainable agriculture is a crucial axis of development for the future of society, which is fully consistent with the objectives of a French bill on to the transition towards sustainable development. This could result in prohibition of pesticides in green public spaces starting in 2017. In 2013, the European Commission restricted the use of three pesticides of the neonicotinoids family responsible for the mortality of bees. Under growing pressure from the public and scientists, numerous initiatives have emerged that press for the withdrawal of glyphosate (RoundUp), whose products of degradation will probably be revealed to be highly carcinogenic.

At this time, bio-control techniques are particularly effective and used in vegetable crops, fruit arboriculture, vineyard and in large-scale farming (see Chapter 1). They are mostly the fruit of research in chemical ecology and more particularly in the classification of the phenomena of attraction, repulsion or induction, in trophic or non-trophic interactions. Bio-control products, preferring the use of natural mechanisms and interactions, are for commercial reasons classified by nature:

– auxiliary macroorganisms of crop protection are invertebrates (insects, mites, nematodes), predator or parasitoids, used to protect cultures against the attacks of biological pests (e.g. inundative releases, autocidal control);

– microorganisms are used to fight against the pathogens responsible for diseases or against insect pests. Entomo-pathogen fungi, for example, play an important role in the natural regulation of insect populations and numerous myco-insecticides and myco-miticides are commercialized;

– chemical mediators comprise the insect pheromones and kairomones (see Chapter 1). They are used to control insect populations. Synthetic molecules, analogs of natural pheromones, are used to perturb the reproduction (sexual confusion) or trap (pheromone traps) harmful insects or pests. Anti-oviposition pheromones will serve as a nesting repellent for phytophage insects. Kairomones serve as food bait to trap pests;

– the natural substances used as products of bio-control are composed of substances present in the natural environment. They can be of vegetable, animal or mineral origin. It is in general the repellents which keep pests at a distance from cultures or insect carriers of disease at a distance from humans and thus reduce the man–carrier contact.

This classification, established on the basis of observations at the macroscopic scale, only takes into account the phenomena of chemical mediation at the molecular scale and the biochemical mechanisms involved. For example, how to classify compounds as elicitors which will activate the natural defense mechanisms of plants? The knowledge acquired in recent years in chemical ecology, by highlighting species-specific relations, the structural characterization and mode of action (attractants, repellents, elicitors, stimulants, etc.) of chemical mediators, have lead researchers involved in bio-control to propose a classification based on the mode of action of the chemical mediators:

– direct-acting compounds including pheromones and allelochemicals, toxic compounds, fungicides and phytotoxic compounds;

– indirect-acting substances, such as the substances stimulating the natural defenses of plants, with compounds mimicking the patterns present on the surface of phytopathogens and involved in the mechanisms of recognition of the aggressor by the plant, and the analog compounds of key vegetable metabolites in the signaling of the aggression or in the awakening of the defense.

Bio-control is a rapidly growing area. Researchers and entrepreneurs work together to provide the conceptual and technological basis to give credibility to and rationalize the development and marketing of new products of crop protection.

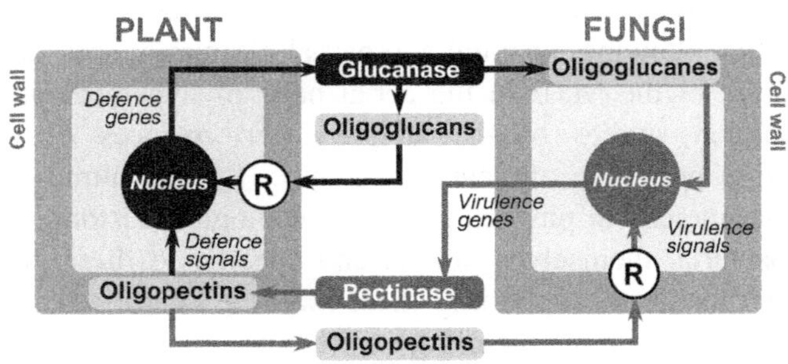

Figure 9.7. *Elicitor and plant–pathogen relations: the example of a fungal attack (http://www.phytopole.com/UMR1931.htm)*

For example, a new biopesticide, Vacciplant®, has recently been introduced to the market. Research on its active principle, laminarin (an oligosaccharide) highlighted the elicitor role of plant defenses.

9.3.2. *Bio-inspired chemistry and remedial phytotechnologies*

Intensive mining and metallurgical industrial activities are causing significant soil contamination by metal species. This is a serious problem because soil performs essential functions which determine food production and water quality. Furthermore, metallic trace elements (MTE) are among the most harmful species and are not biodegradable.

Beyond environmental consequences, the health risks are real: nervous system, kidney, pulmonary or bone tissue damage is clearly established. Recent examples of blood lead levels and the first stage of lead poisoning have been highlighted in children living near old mine sites. Specific impacts of metallic pollution are not only of an environmental and medical order, but they also directly affect the economic and tourism development of the areas concerned. The excavation or confinement of contaminated areas has been set up in certain sites, but these techniques cannot provide satisfying solutions. Being uninspired and poorly accepted, they are also expensive and resolve nothing.

A few years ago, rare plants were discovered which are capable of growing on desert soils which had become phytotoxic. These plants progressively developed adaptation strategies, helping them to tolerate the pollution and even extract the MTE in order to store them in their aerial parts so as to better protect themselves. The amounts of MTE found in the leaf systems can reach impressive levels (up to 7–8% of dry mass). This is called phytoextraction. It is an ecotechnology allowing a natural and partial cleanup of soils and sediments by accumulation of the MTE in the aerial parts of hyperaccumulator plants. Recent evaluation studies of the adaptive performances of these plants highlighted the presence of leguminous-type hyperaccumulating species, reinforcing the interest, in phytoextraction in ecological restoration programs. Associated bacteria, themselves unique and specific to these polluted sites, have become capable of supporting these extreme conditions. They behave like a factory by assimilating the nitrogen present in the air to transform it into a natural fertilizer to aid the growth of other plants. In exchange, the plants produce carbonated nutrients by photosynthesis destined for bacteria, in an impoverished soil. Despite the study and understanding of these ingenious natural systems, the development of phytoextraction has been limited by absence of valorization of the contaminated biomass, which was considered as dangerous and useless.

Recently, innovative processes of valorization of these extraordinary plants have emerged [ESC 14]. They are the basis of a new domain of green chemistry, called ecocatalysis. Taking advantage of the remarkable adaptive capacity of these plants to hyperaccumulate the Zn^{2+}, Ni^{2+}, Mn^{2+}, Cu^{2+}, Co^{2+} and Pd^{2+} cations, the conception of the program is based on the direct use of the plant-sourced metallic species as "Lewis acids", catalysts for organic chemical reactions. The principle rests on an approach mimicking natural catalysts, enzymes. They help accelerate infinitely slow reactions and are regenerated after transformation of the organic molecules. Catalysts produced from the hyperaccumulating plants are capable of catalyzing more complex reactions. They consist of original and rare chemical species, that metallurgy has never been able to produce. These biosourced catalysts show unusual reactivity. Often superior to classic catalysts from chemistry, they allow the synthesis of molecules of interest (anticancer agents, antiviral agents, molecules active against malaria, natural scents, biocosmetics, new generation and bio-inspired insecticides and key intermediates of the chemical industry) under mild and surprisingly effective conditions. The synthesis strategies rely on ecocatalyzed biomimetic conceptions of green reduction (analogs of NADPH catalyzed by ecocatalyze) and from green oxidation (analogs of manganese porphyrins) of cytochrome P450.

9.3.2.1. *Example of biomimetic reduction*

The NADPH is an enzymatic cofactor, and more precisely a source of hydride – namely the H^- anion which serves to reduce chemical functions. The principle of this reduction is illustrated below (Figure 9.8) in the case of the reduction of the pyruvic acid into lactic acid.

A chemical reduction, also based on a hydride transfer, was also developed: the amino-reduction reaction. One example of such a reaction applied to a bioactive molecule is presented below (Figure 9.9).

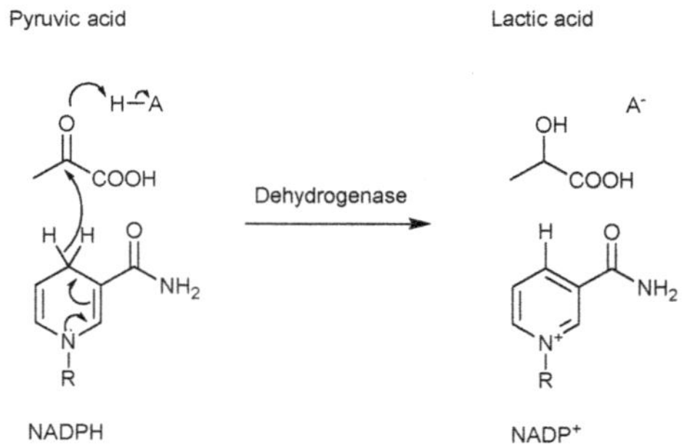

This reaction occurs in the muscle cells during intense activity and in an anaerobic environment. This biological process is the starting point of a new method of biomimetic reduction. It is based on the use of a synthetic analog of NADPH: the Hantzsch ester, or diludin. The catalytic effect of the enzyme is replaced by the use of a biosourced ecocatalyst rich in manganese (Eco-Mn®).

Figure 9.8. *Reduction of pyruvic acid into lactic acid*

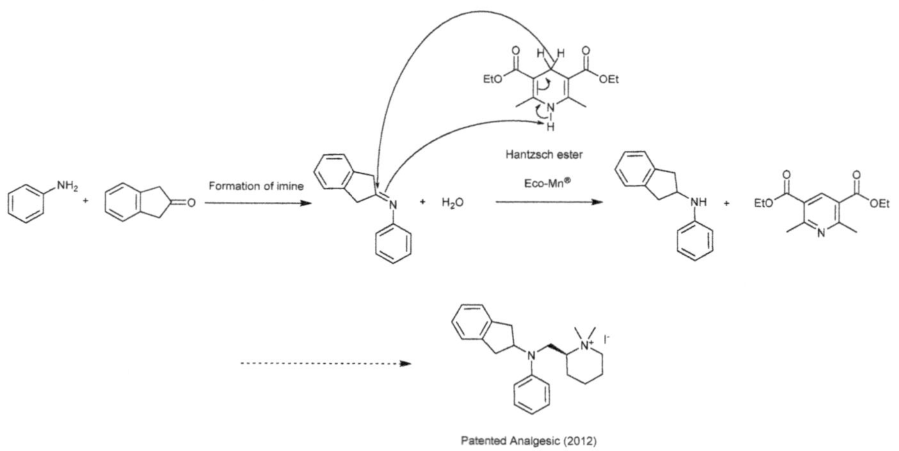

Figure 9.9. *Reductive animation catalyzed by Eco-Mn® by using a Hantzsch ester as a hydride donor*

This methodology allowed us, for example, to synthesize the product of the amino-reduction of aniline with 2-indanone to obtain the precursor of a series of new molecules helping to treat pain in patients. It was also possible to demonstrate all the utility of a biosourced catalyst, including in the field of the pharmaceutical industry. This methodology has been recently protected by a patent [GRI 14].

9.3.2.2. *Example of biomimetic oxidation*

In the living world, there is an entire class of enzymes charged with oxidizing organic compounds to facilitate their elimination: the cytochrome P450. Within the active site of these enzymes is found the heme, consisting of a porphyrin derivative complexing in it an iron atom. This heme is an essential part of the cytochrome P450; the iron atom, protected within cytochromes, has the function to activate an oxygen molecule destined to transfer one of its atoms to the molecule needing to be oxidized.

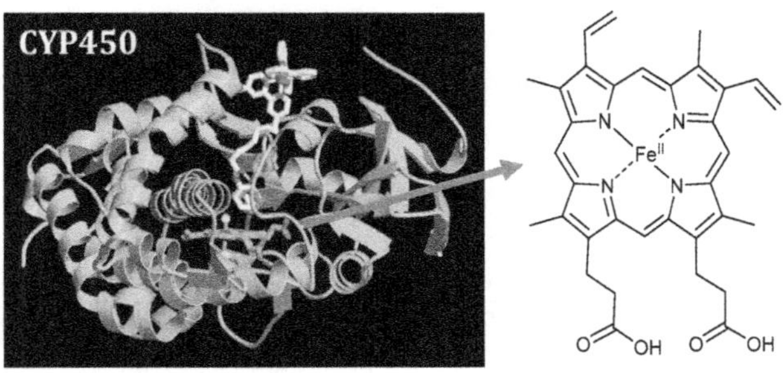

Figure 9.10. *Ribbon model of cytochrome P450 with heme*

The understanding of biological phenomena at the molecular scale is very complex. It involves chemical reactions identical to those classically known by chemists, but the tridimensional arrangement of the reagent is much more sophisticated. A fine mechanistic understanding of these biological systems helps to conceive simplified copies of them. By receiving inspiration from biological oxidants and their geometric organization in the cytochrome P450, analog of natural hemes has been prepared in order to achieve neighboring oxidation reactions in mild conditions [GRI 14]. The goal is to successfully and effectively prepare, by green processes, new oxidation reagents capable of replacing the harmful oxidants faulted by European regulations of chemicals REACH. The basic structure of porphyrin heme was conserved, while simplifying it, and the Iron(II) cation was replaced by Eco-Mn® enriched in manganese in oxidation state III. Oxidations of alkenes were finally achieved in order to convert them to epoxide (Figure 9.11).

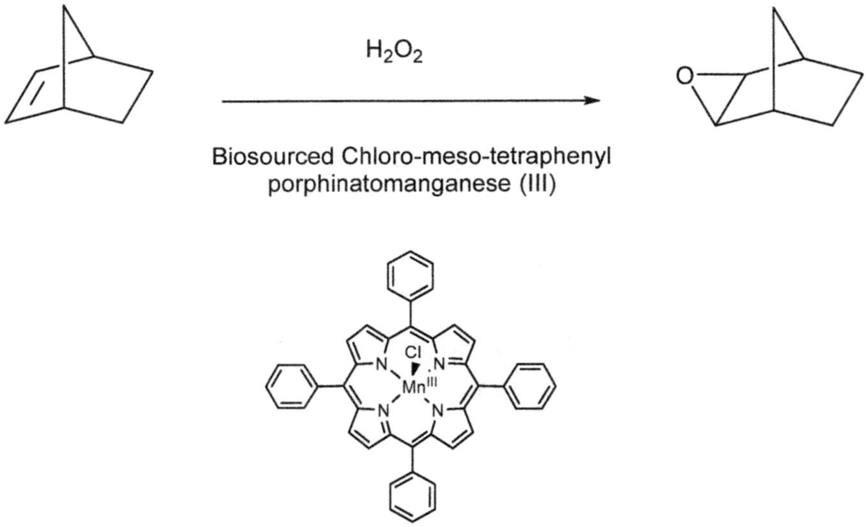

Figure 9.11. *Epoxidation of norbornene by oxygenated water, catalyzed by a biosourced porphyrin of manganese*

In the margins of the examples presented above, many of the prepared molecules are present in nature. The chemist is not responsible for their biological activity; he just observes, tries to understand, explain and be inspired by it. The synthesis of these molecules helps however to preserve natural resources.

Nature has, therefore, found solutions to adapt to heavy pollution generated by anthropic activities: phytoextraction. The discovery and the understanding of this natural system is today the source of vast ecological restoration programs at numerous mining sites (France, New Caledonia, China, Gabon, etc.). That also helps to develop innovative solutions for developing recycling of mineral resources and constitutes a concrete solution to the criticality of non-renewable minerals. Finally, these plants help to develop innovative chemical catalysts of unequalled performance.

This work of interdisciplinary research is completely bio-inspired. In applied and industrial purposes, it intends to be a driver of environmental and socio-economic reconstruction of sites marked by industrial and mining activities.

9.4. Conclusions

The performances of nature are unique: generators of precious chemical systems and adaptable to extreme conditions. It is a multidisciplinary source of inspiration, which the researcher must understand in order to sustainably valorize and restore it if necessary. Nature can be the starting point of a new circular green economy sector that reconciles ecology and chemistry. The characterization of chemical mediation, and therefore functional metabolites, selected over thousands of years of evolution and co-evolution for their efficacy and selectivity on cellular receptors, feeds research domains involved in medicinal chemistry, pharmacology, phytopharmacy or on a wider scale in chemistry for everyday life.

9.5. Bibliography

[BAN 14] BANAIGS B., BONNARD I., WITCZAK A. *et al.*, "Marine peptides secondary metabolites, in outstanding marine molecules and new trends in analytical methods", in LA BARRE S., KORNPROBST J.M. (eds), *Outstanding Marine Molecules*, Wiley-VCH, Weinheim, 2014.

[BED 14] BEDOUX G., HARDOUIN K., MARTY C. *et al.*, "Chemical characterization and photoprotective activity measurement of extracts from the red macroalga *Solieriachordalis*", *Botanica Marina*, vol. 57, pp. 291–301, 2014.

[BEN 14] BEN HAJ SALAH K., INGUIMBERT N., "Efficient microwave-assisted one shot synthesis of peptaibols using inexpensive coupling reagents", *Organic Letters*, vol. 16, pp. 1783–1785, 2014.

[BEN 98] BENYUS J.M., *Biomimicry: Innovation Inspired by Nature*, HarperCollins Publishers, 1998.

[BER 15] BERTRAND C. (ed), *Natural Product and Biocontrol*, Presses Universitaires de Perpignan, 2015.

[BES 06] BESNARD O., MARTINEZ J., CAVELIER F., Oligopeptides, composition and use thereof as elicitors of the natural defences of plants, Patent US20060148710 A1, 2006.

[EBR 13] EBRIGHT R.H., SEVERINOV K., Nucleic acid sequences for biosynthesis of non-MccJ25-related lariat peptides, Patent US8461314, 2013.

[ESC 14] ESCANDE V., GAROUX L., GRISON C. *et al.*, "Ecological catalysis and phytoextraction: Symbiosis for future", *Applied Catalysis B-Environmental*, vol. 146, pp. 279–288, 2014.

[GRI 14] GRISON C., ESCANDE V., Utilisation de certaines plantes accumulatrices de manganèse pour la mise en œuvre de réactions de chimie organique, Patent WO2014016509 A1, 2014.

[HON 11] HONG J., "Role of natural product diversity in chemical biology", *Current Opinion in Chemical Biology*, vol. 15, pp. 350–354, 2011.

[KAL 15] KALIA V.C. (ed.), *Quorum Sensing vs Quorum Quenching: A Battle with No End in Sight*, Springer, 2015.

[LI 14] LI Y., ZIRAH S., REBUFFAT S., *Lasso Peptides: Bacterial Strategies to Make and Maintain Entangled Scaffolds*, Springer-Verlag, 2014.

[PIS 15] PISCOTTA F.J., THARP J.M., LIU W.R. *et al.*, "Expanding the chemical diversity of lasso peptide MccJ25 with genetically encoded noncanonical amino acids", *Chemical Communications*, vol. 51, pp. 409–412, 2015.

[POU 11] POUPON E., NAY B. (eds), *Biomimetic Organic Synthesis*, Wiley-VCH, 2011.

[ROB 98] ROBERTS M.F., WINKS M. (eds), *Alkaloids, Biochemistry, Ecology and Medicinal Application*, Plenum Press, New York, 1998.

[SAN 13] SANCHEZ L.M., DORRESTEIN P.C., "Analytical chemistry: Virulence caught green-handed", *Nature Chemistry*, vol. 5, pp. 155–157, 2013.

[SIN 97] SIN N., MENG L., WANG M.Q.W. *et al.*, "The anti-angiogenic agent fumagillin covalently binds and inhibits the methionine aminopeptidase, MetAP-2", *Proceedings of the National Academy of Sciences*, vol. 94, pp. 6099–6103, 1997.

[YU 11] YU M.J., ZHENG W.J., SELETSKY B.M. *et al.*, "Case history: Discovery of eribulin (HALAVEN™), a halichondrin B analogue that prolongs overall survival in patients with metastatic breast cancer", *Annual Reports in Medicinal Chemistry*, vol. 46, pp. 227–241, 2011.

[WAN 07] WANG G., SHANG L., BURGETT A.W.G. *et al.*, "Diazonamide toxins reveal an unexpected function for ornithine δ-amino transferase in mitotic cell division", *Proceedings of the National Academy of Sciences*, vol. 104, pp. 2068–2073, 2007.

Conclusion

Looking Forward: the Chemical Ecology of Tomorrow

Biochemistry, biophysics, astrobiology, astrophysics, etc. – quite often the exchanges and reconciliations between scientific disciplines whose objects of study and paradigms are sometimes quite distant from each other have proven extraordinarily fruitful and have even helped give birth to new disciplines. Chemical ecology, born of a convergence between chemistry and ecology, is thus a particularly successful illustration of the interest of pursuing these interdisciplinary dialogs. It offers a framework that helps to advance our understanding of chemobiodiversity and of the chemical language of nature, along a continuum from the gene to the ecosystem. This book testifies to the great progress researchers have made in understanding the role of chemical mediation in signaling and defenses in very diverse biotopes, aquatic or terrestrial, and between organisms in the animal, plant, fungal or bacterial kingdoms. Chemical ecology has become a source of inspiration for new applications and has spurred the conception of future ecotechnologies that will make essential contributions to the resolution of environmental problems.

The resolutely interdisciplinary approach that chemical ecology has followed in describing and understanding this invisible communication must be pursued, because we still know very little about chemical mediation in numerous lineages of organisms, both terrestrial and marine. Particular efforts are still necessary to understand sometimes highly complex networks of interactions, to allow a more global comprehension of evolutionary

---

Conclusion written by Martine HOSSAERT-MCKEY and Anne-Geneviève BAGNÈRES.

mechanisms, and even to help devise effective conservation strategies for the partners in these interactions. The analysis of chemical mediation in the context of the evolutionary dynamics of interactions among species will have to take into account not only phylogenetic inertia, which can constrain the synthesis and construction of olfactory messages, but also factors conferring flexibility, such as phenotypic plasticity or even epigenetic modification. Finally, in the years to come, the use of genomic, transcriptomic and other "omics" approaches will enable the parallel analysis of emitter and receptor genes (have they co-evolved?) in order to reveal an image in which the often complex mechanisms behind the chemical interactions among living organisms are integrated with evolutionary genetics.

Our efforts to understand these interactions and the mechanisms underlying them must be especially reinforced in tropical environments, where chemobiodiversity "explodes" and where the games of interspecies interactions become particularly complex. A new and still rarely explored dimension is that of chemical mediation at the scale of entire biotic communities, involving a more integrated approach and including the important spatial dimension.

One of the significant advances in ecology over the past 20 years has been the demonstration of synergies between different pressures and environmental changes affecting the functioning and evolution of ecosystems. The chemoecological dimensions of these synergies are yet to be explored. For example, in marine environments, where chemicals diffuse rapidly and extensively and where chemical mediation plays many important roles, a major focus of current research is to better understand how global environmental changes could modify the networks of chemically mediated interactions that influence the processes of biological invasions. In both marine and terrestrial systems, in both natural environments and intensively managed agroecosystems, studies of how these changes impact interactions among species and the structuring of communities are just beginning.

In parallel, chemical ecologists are participating in the development of ecogeochemistry, an emerging domain at the interface of chemistry and ecology that combines approaches centered on mediation among organisms – the traditional province of chemical ecology – with approaches centered on the interactions of organisms with the physicochemical components of their environments. The major challenge of ecogeochemistry is the development of tools and experimental strategies to help us understand the fundamental mechanisms of species dynamics in the context of their environment. The objective in the years to come is, therefore, to predict the dynamics of the networks of interactions that link diversity of the living to the physicochemical components of the environment, to predict not only the effects of environmental changes on biodiversity, but also the effects of changes in biodiversity on the biogeochemical functioning of environments. Developing such a dialectical approach is crucial for understanding the resilience capacity of ecosystems. One future application is the definition of protocols in ecological engineering, for example, for the remediation of polluted environments.

Among the major advances in chemical ecology presented in this work, research in two domains appears particularly original: new explorations of chemical ecology in primate societies and in microbial communities. The chemical ecology of primates, both non-human and human, offers numerous perspectives in multiple research themes. We can cite, for example, analysis of the emission of odor signals, of their reception and their role in regulating interindividual relations and structuring populations; the individual and collective acquisition of therapeutic expertise; the techniques of chemical defense employed against parasites or predators of humans or of their animal or plant resources; the olfactory techniques of body care, the fabrication of olfactory "ornaments" and the production of odors to create domestic atmospheres and settings for ritual practices; and the management of chemospheres and hydrospheres of the urban centers where human populations of the future will increasingly reside.

The discovery during the last 30 years of an extraordinary diversity of microorganisms, inhabiting environments that we previously believed hostile to life, has opened new perspectives not only for the study of functionality and stability of biological macromolecules but also for investigating the origins of life on earth and on other planets. These "extremophile" microorganisms, which include not only bacteria but also archaea, which constitute one of the three domains of living things (Bacteria, Archaea and Eukarya), are often associated in communities living in biofilms, where they require intra- and interspecies communication and mechanisms to regulate cell densities, such as quorum sensing (QS). The study of QS in these microorganisms is an expanding domain of research, with spillovers in biotechnology. In addition, the advent of new techniques in spectroscopy (high-field NMR) and molecular analysis (metabolomics and genomics), associated with the expansion of *in situ* imagery and integrative methods of studying the holobiont (metabolic networks in a complex of associated organisms), will help us better track chemical mediators that are often produced not only in very small quantities, but also under specific conditions and that are subject to numerous sources of variation. The notion of the holobiont is now widespread and could be generalized to a large number of organisms and metabolomic techniques, contributing to our understanding of the complexity of numerous organisms living in association. Likewise, as witnessed in recent advances in "omics" techniques, a large number of chemical mediators involved in communication among microorganisms are not expressed under standard laboratory conditions, suggesting a chemical diversity yet more vast than what has already been discovered. The classification of these new compounds and the determination of their ecological roles will undoubtedly accelerate the development of new compounds of high added value with applications in medicine, ecology, agronomy and even biotechnology.

Despite the significant advances illustrated in this book, methodological integration – the pillar of chemical ecology – still requires reinforcement in laboratories and research teams in this domain. It is urgent to capitalize our efforts by investing in equipment and research programs (funded by national and international research agencies, notably European) dedicated to chemical ecology and to metabolomics. At an experimental level, developing mesocosms and other infrastructures for conducting experimental studies of chemical mediation is a promising pathway for progress in decrypting these complex processes, whose study is still more often confined to simple laboratory conditions or to uncontrolled field observations. In another domain, the use in chemical ecology of technologies such as NGS (next-generation sequencing) and ultra-high-performance analytical chemistry has so far mobilized only a relatively small number of researchers, a new community at the forefront of these advanced techniques in genomics, transcriptomics and metabolomics. We are currently witnessing a real technological revolution, which has occurred very quickly and which requires the scientific community to adapt quickly to it. For researchers to meet this major challenge, it will be important to provide training in new domains (e.g. in bioinformatics for the processing of increasingly complex data), to promote the transfer of competence between teams (through workshops, thematic schools, research consortia, etc.) and to provide structure for shared use (access to sequencing platforms, technical personnel, etc.).

In chemical ecology, as in many other domains, technological progress has accelerated the elaboration of numerous databases, with the principal goal of enhancing collaboration. Ironically, however, the first result of this progress is a fragmentation of information through the creation of multiple small bases. Centralizing data in larger, more integrative databases would make collaboration easier and would also ensure the permanence of data. An effort should be made to achieve standardization of these multiple databases, and even to constitute bases of metadata in chemical ecology, as has already been done for genomic data, for example. A second way in which technological progress has ironically produced a new obstacle is that progress in "omics" has produced enormous quantities of data, beyond our present capacity to process them. To deal with the flow of information and to interpret this information in a pertinent manner, effective bio-statistical tools must be developed.

Today, more than ever before, scientific and technological advances involve work at the frontiers of different disciplines. Nowhere is this truer than in work on bio-inspired technologies. In this domain, at the frontiers of biology, ecology and chemistry, we must understand how the functioning of living systems, often very complex, translates into new ecologies or into new chemistries. Researchers in chemical ecology are well prepared for such work, as they navigate constantly between biological models and chemical mediators, between chemical mediators and sensory receptors, and between fundamental research and research applied to developing new technologies.

Glossary

Allelopathy: any direct or indirect harmful or beneficial effects of one plant (including microorganisms) upon another through production of chemical compounds that escape into environment. The field of allelopathy has broadened over the years and now addresses research in both terrestrial (natural and agricultural) and aquatic (marine and riverine) ecosystems.

Beta diversity: measure of biodiversity, which consists of comparing the diversity of species among different ecosystems or along environmental gradients.

Chemical mediator (or semiochemical or infochemical): chemical substance emitted in the environment, which has a chemical signal value among one or more organisms.

DNA microarrays: collection of DNA molecules fixed in an ordered array on a small surface (glass, silicon or plastic), which are used as probes in a pair of complementary strands of DNA contained in a biological extract and thus reveal the level of expression of the genes (transcripts) in a cell, tissue, organ or organism.

Expressed Sequence Tag (EST): short sequenced portion of a complementary DNA (cDNA), used as a marker to differentiate the genes among those in a DNA sequence.

Genomics: study of the genomes of organisms, concerned with the sequencing and analyses of DNA sequences present in organisms.

Growth–defense trade-off: compromise between the allocation of resources allocated to growth and those allocated to defense.

Herbivore-induced plant volatiles (HIPVs): volatile compounds produced by plants in response to herbivore attacks.

Quorum sensing: synchronization mechanism of a population of bacteria. Bacteria that use quorum sensing exchange small molecules as a chemical mediator in order to modulate their behavior. Once the quorum sensing is established, bacteria start to act as multicellular organisms and form real biofilms.

Latex (natural): emulsion produced in the latex cells composed of polymer particles (polylactide or polyisoprene) in an aqueous medium.

Metabolomics: study of metabolites in living organisms at a large scale. It allows the identification and analyses of all metabolites (chemical components) constituting the metabolome of cells, organs, or organisms, in given physiological conditions or environments.

Proteomics: study of proteins in living organisms at a large scale. It allows the identification and analyses of all of the proteins constituting the proteome of organisms, organs or cells in given physiological conditions or environments.

Restriction-site-associated DNA (RAD-seq): short fragments of DNA adjacent to DNA sequences which could be cut by specific restriction enzymes. The use of RAD markers helps to analyze the polymorphism of individuals.

RNA sequencing (RNA-Seq): sequencing technique of the entire transcriptome (all the RNA sequences) of a species. This technique allows the identification and analysis of the expression profiles of the collection of transcripts from a given tissue, organ or organism. This technique is particularly effective for non-model species (e.g. those without a reference genome).

Single nucleotide polymorphism (SNP): polymorphism of the DNA sequence for which the two homologous chromosomes of an individual show a difference in a single base pair. An SNP can be detected between homologous chromosomes in a single individual or between individuals.

Transcriptomics: study of gene expression at a large scale. It allows the identification and analyses of all transcripts (messenger RNAs) constituting the transcriptome of cells, organs, or organisms, in given physiological conditions or environments.

Volatile organic compounds (VOC): organic gases, from human-made or natural sources, which can propagate a greater or lesser distance from their place of emission in the atmosphere (or other medium) and can result in direct or indirect effects on organisms.

List of Authors

Ali AL MOURABIT
UPR 2301
Institut de Chimie des Substances Naturelles
CNRS, University Paris-Sud
University Paris-Saclay
Gif-Sur-Yvette
France

Anne-Geneviève BAGNÈRES
UMR 7261
Institut de Recherche sur la Biologie de l'Insecte
CNRS, University of Tours
France

Virginie BALDY
UMR 7263
Institut Méditerranéen de Biodiversité et d'Ecologie Marine et Continentale (IMBE)
Aix Marseille University
University of Avignon
CNRS, IRD
Marseille
France

Bernard BANAIGS
USR 3278
Centre de Recherche Insulaire et Observatoire de l'Environnement
CNRS, EPHE
University of Perpignan
France

Nicolas BARTHÈS
UMR 5175
Centre d'Ecologie Fonctionnelle et Evolutive
CNRS
Paul Valéry Montpellier University
EPHE
Montpellier
France

Sylvie BAUDINO
FRE 3727
Laboratoire de Biotechnologies Végétales appliquées aux Plantes Aromatiques et Médicinales
CNRS, University of St-Etienne
University of Lyon
France

Jean-Claude CAISSARD
FRE 3727
Laboratoire de Biotechnologies Végétales appliquées aux Plantes Aromatiques et Médicinales
CNRS, University of St-Etienne
University of Lyon
France

Marie CHARPENTIER
UMR 5175
Centre d'Ecologie Fonctionnelle et Evolutive
CNRS
Paul Valéry Montpellier University
EPHE
Montpellier
France

Guillaume CLAVÉ
FRE 3673
Laboratoire de Chimie bio-inspirée et d'Innovations écologiques
CNRS, University of Montpellier
France

Anne-Marie CORTESERO
UMR 1349
Institut de Génétique, Environnement et Protection des plantes
Inra, Agrocampus Ouest
University of Rennes 1
France

Christophe DUPLAIS
UMR 8172
Ecologie des Forêts de Guyane
AgroParisTech, Cirad, CNRS, Inra, University of the French West Indies and Guiana
Cayenne
Guiana

Catherine FERNANDEZ
UMR 7263
Institut Méditerranéen de Biodiversité et d'Ecologie marine et continentale (IMBE)
Aix Marseille University
University of Avignon
CNRS, IRD
Marseille
France

Xavier FERNANDEZ
UMR 7272
Institut de Chimie de Nice
CNRS
University Nice Sophia Antipolis
France

Claude GRISON
FRE 3673
Laboratoire de Chimie bio-inspirée et d'Innovations écologiques
CNRS, University of Montpellier
Montpellier
France

Martine HOSSAERT-MCKEY
UMR 5175
Centre d'Ecologie Fonctionnelle et Evolutive
CNRS
Paul Valéry Montpellier University
EPHE
France

Jérémy JUST
UMR 5667
Laboratoire Reproduction et Développement des Plantes
ENS Lyon, INRA, CNRS
University Lyon 1
France

Catherine LEBLANC
UMR 8227
Laboratoire de Biologie Intégrative des Modèles Marins-Station Biologique de Roscoff
CNRS
Pierre and Marie Curie University
Roscoff
France

Nadine LE BRIS
UMR 8222
Laboratoire d'Ecogéochimie des Environnements Benthiques
UPMC-CNRS
Banyuls
France

Christophe LUCAS
UMR 7261
Institut de Recherche sur la Biologie de l'Insecte
CNRS University of Tours
France

Doyle MCKEY
UMR 5175
Centre d'Ecologie Fonctionnelle et Evolutive
CNRS
Paul Valéry Montpellier University
EPHE
Montpellier
France

Florence NICOLÈ
EA 3061
Laboratoire de Biotechnologies Végétales appliquées aux Plantes Aromatiques et Médicinales
University of St Etienne
France

Guillaume ODONNE
USR 3456
Laboratoire Ecologie, Evolution, Interactions des Systèmes Amazoniens
CNRS
University of Guyane
IFREMER
Cayenne
French Guiana

Thierry PÉREZ
UMR 7263
Institut Méditerranéen de Biodiversité et d'Ecologie marine et continentale (IMBE)
Aix Marseille University
University of Avignon
CNRS, IRD Marseille
France

Philippe POTIN
UMR 8227
Laboratoire de Biologie Intégrative des Modèles Marins-Station Biologique de Roscoff
CNRS
Pierre and Marie Curie University
Roscoff
France

Soizic PRADO
UMR 7245
Molécules de Communication et Adaptation des Micro-organismes
MNHN – CNRS
Paris
France

Magali PROFFIT
UMR 5175
Centre d'Ecologie Fonctionnelle et Evolutive
CNRS
Paul Valéry Montpellier University
EPHE
Montpellier
France

Sylvie REBUFFAT
UMR 7245
Molécules de Communication et Adaptation des Micro-organismes
MNHN
CNRS
Paris
France

Benoist SCHAAL
UMR 6265 Centre des Sciences du Goût et de l'Alimentation
CNRS, University of Burgundy
INRA
Dijon
France

Bertrand SCHATZ
UMR 5175
Centre d'Ecologie Fonctionnelle et Evolutive
CNRS
Paul Valéry Montpellier University
EPHE
Montpellier
France

Carole SMADJA
UMR 5554
Institut des Sciences de l'Evolution de Montpellier
CNRS, IRD
Montpellier University 2
Montpellier
France

Olivier P. THOMAS
UMR 7272
Institut de Chimie de Nice
CNRS
University Nice Sophia-Antipolis
Nice
France

Frédérique VIARD
UMR 7144
Adaptation et Diversité en Milieu Marin
Station biologique de Roscoff
CNRS, UPMC
Roscoff
France

Index

A, B, C

D, E, F

G, H, I

L, M, N

O, P, Q

R, S

T, U, V

CPSIA information can be obtained at www.ICGtesting.com
Printed in the USA
BVOW06*0933040816

457873BV00001B/3/P

9 781848 219243